Prentice Hall
4/8/85
30.00

THE ENGINEERING
OF NUMERICAL SOFTWARE

PRENTICE-HALL SERIES IN COMPUTATIONAL MATHEMATICS
Cleve Moler, advisor

THE ENGINEERING OF NUMERICAL SOFTWARE

WEBB MILLER

The University of Arizona

PRENTICE-HALL, INC.
Englewood Cliffs, New Jersey 07632

Library of Congress Cataloging in Publication Data

Miller, Webb.
The engineering of numerical software.

Bibliography: p.
Includes index.
1. Numerical analysis—Computer programs. I. Title.
QA297.M527   1984   519.5   84-9830
ISBN 0-13-279043-2

Editorial / production supervision: Nancy Milnamow
Cover design:
Manufacturing buyer: Gordon Osbourne

093496

© 1984 by Prentice-Hall, Inc., Englewood Cliffs, New Jersey 07632

All rights reserved. No part of this book may be
reproduced, in any form or by any means,
without permission in writing from the publisher.

Printed in the United States of America

10 9 8 7 6 5 4 3 2 1

0-13-279043-2

Prentice-Hall International, Inc., *London*
Prentice-Hall of Australia Pty. Limited, *Sydney*
Editora Prentice-Hall do Brasil, Ltda., *Rio de Janeiro*
Prentice-Hall Canada Inc., *Toronto*
Prentice-Hall of India Private Limited, *New Delhi*
Prentice-Hall of Japan, Inc., *Tokyo*
Prentice-Hall of Southeast Asia Pte. Ltd., *Singapore*
Whitehall Books Limited, *Wellington, New Zealand*

CONTENTS

PREFACE		**vii**
Chapter 1. INTRODUCTION		**1**
1.1	Highlights of the Software Engineering Process	*1*
1.2	An Example of a Test and Two Experiments	*6*
1.3	An Example of Performance Measurement	*11*
1.4	Our Choice of Topics	*14*
Chapter 2. FLOATING-POINT ARITHMETIC		**17**
2.1	Our Assumptions About Floating-Point Arithmetic	*18*
2.2	Parameterizing Numerical Software	*20*
2.3	Ulps	*32*
2.4	Uncertainty Diagrams	*36*
Chapter 3. COMPUTATION OF THE SINE FUNCTION		**45**
3.1	A Rudimentary Sine Procedure	*47*
3.2	Implementation Issues: Small and Large Angles	*52*
3.3	Testing Your Sine Procedure	*54*
3.4	A More Accurate Sine Procedure	*61*
3.5	Testing High-Accuracy Sine Procedures	*65*

Chapter 4. LINEAR EQUATIONS 72

4.1 Elementary Facts About Linear Equations 74
4.2 Implementation Details 81
4.3 Testing a Linear Equation Solver 93
4.4 A Property that Almost Always Holds 107

Chapter 5. SOLVING A NONLINEAR EQUATION 109

5.1 Bisection 110
5.2 Incorporating Linear Interpolation 114
5.3 The Importance of Performance Measurements 119
5.4 Minimization (Optional) 128

Chapter 6. INTEGRATION 138

6.1 Simpson's Rule 139
6.2 A Simple Procedure for Automatic Integration 144
6.3 Measuring the Performance of Integration Procedures 150
6.4 Sturdier Cost-Versus-Error Graphs 155

INDEX 165

PREFACE

A number of important and interesting concepts about numerical software began to emerge in the 1970s: concepts that are independent of traditional numerical analysis and possess intrinsic merit and appeal. At the same time, experience gathered by publicly funded research projects and by the development of commercial software libraries raised understanding of the broad issues of numerical software production to the point that those concepts can be woven into a body of knowledge appropriate for systematic study.

This book is my attempt to make these advancements readily accessible. Whereas books such as *Computer Solution of Mathematical Problems* by George Forsythe, Michael Malcolm, and Cleve Moler (Prentice-Hall, Englewood Cliffs, N.J., 1977) teach the *use* of numerical software, much as introductory programming courses teach students to use a compiler without learning how to build one, my goal is to present principles for *writing* numerical software. The ideal textbook about the production of numerical software remains to be written, but I hope that I have verified its worth and feasibility and hastened its arrival.

The book is primarily intended for use as the text in a one-term course for students of computer science, engineering, science, and mathematics. It can also be used for self-study or for supplementary reading in a more traditional course on numerical computation. The material is accessible to readers with one year's programming experience, a minimum of training in mathematics, and the general level of academic maturity that can be expected of a senior undergraduate or first-year graduate student.

I believe that computer scientists will find this material useful, understandable, and even enjoyable. Educated guesses attribute as much as 50% of all software costs to numerical software, and it is appropriate that a study of computer science include the issues particular to that area.

Scientists, engineers, and mathematicians can also profit from exposure to this material. In some situations an awareness of the tremendous effort required to produce numerical software will lead them to use packaged software. In cases where software is not available and is worth the cost of development, a thorough understanding of the ingredients that go into successful software for several computational problems provides insight into what must be done for the problem at hand.

The topics covered, and their order of presentation, were chosen to proceed systematically across a spectrum of numerical software. There is, however, some flexibility in the order that the material is studied: Chapters 3–5 can be taken in any order once Chapters 1 and 2 have been read, while Chapter 6 can immediately follow Chapter 1.

The programming assignments and exercises are integral parts of the text. I recommend that the reader write all the programs and work all the exercises, except for material that is explicitly labeled as optional.

FORTRAN 77 programs are sprinkled throughout the book, so a reading knowledge of the language will be needed. Experience indicates that time will be saved in the long run if the reader invests whatever effort is needed to write the assigned programs in FORTRAN.

I would like to thank the faculty and students of the Department of Computer Science at the University of Arizona for providing the ideal environment for writing this book. Many insightful suggestions were offered by Tim Budd, Helen Deluga, Lee Henderson, James W. Johnson, and Titus Purdin.

Alan George of the University of Waterloo read the entire manuscript and recommended several improvements. My special thanks go to Jim Cody, Jack Dongarra, and James Lyness of the Argonne National Laboratory. Through their writings and direct comments they have contributed immeasurably to this book.

Webb Miller

1
INTRODUCTION

The purpose of this chapter is to explain what the book is all about. We begin, in Section 1.1, by defining some key concepts concerning the process of developing numerical software.

Sections 1.2 and 1.3 illustrate these concepts in the contexts of solving a quadratic equation and checking the solvability of a nonlinear equation. In addition to clarifying our definitions, these examples provide an opportunity for us to point out reasons why the material covered in this book is important. In particular, Section 1.3 illustrates the difficulties associated with fixed-precision arithmetic and motivates our devoting an entire chapter, Chapter 2, to the subject.

Using the vocabulary developed in Sections 1.1–1.3, Section 1.4 outlines the common structure of Chapters 3–6 and explains the systematic differences among them.

1.1 HIGHLIGHTS OF THE SOFTWARE ENGINEERING PROCESS

It is useful to think of the process of producing numerical software as divided into four activities: (1) determining specifications, (2) designing and analyzing algorithms, (3) implementing software, and (4) testing and measuring programs. Of course, the production of an actual program often does not follow an orderly progression through these phases. (For instance, we will see that algorithm analysis may have to precede formulation of a specification.) In addition, we will classify programming mistakes into three loose categories: (1) blunders, (2) numerical oversights, and (3) portability oversights. In spite of being vague, overlapping, and incomplete, this classification facilitates later discussions.

The meanings that we will attach to the terms used in these two categorizations and some related concepts will be explained in this section. In addition, we will describe a class of computational experiments that will be used to investigate the process of locating simple programming mistakes. The reader is urged to consider the definitions carefully but should keep in mind that our definitions are not standard.

Four Phases of Software Production. By a *specification* we mean a precise and complete statement of the intended relationship between a program's input and its computed output. One of the surprising things about numerical software is that for many computational problems it is not possible to determine an appropriate specification, in which case the software must be developed with only an imprecise idea of what it will actually do. Even when specifications exist, they may be difficult to determine. (See Exercise 1.)

An *algorithm* is an outline of the sequence of arithmetic operations that the software is to perform on "unexceptional" data. Much of the activity of designing algorithms is conceptual in nature. Often there are mathematical theorems to be understood and algebraic formulas to be manipulated. The process of algorithm design may include a theoretical analysis of the algorithm's resilience to the use of fixed-precision arithmetic.

The goal of the *implementation* phase of numerical software development is to produce a running computer program. Doing so involves determining the calling sequences, the handling of exceptional cases, the use of storage, etc.

For discussing the software development process, let us distinguish three purposes for running programs on a computer: testing, measuring, and experimenting. (We will not attempt to classify uses of the finished software.) A *test* is conducted on a specification and an associated computer program and is performed by running the program on one or more sets of input data and checking for conformance of the input and output to the specification. On the other hand, we will say that a program is being *measured* if either (1) we have no specification for input/output behavior or (2) we are checking some aspect of program performance other than input/output behavior (for instance, we might be determining efficiency). The act of running a program for the stated purpose of investigating some aspects of the software engineering process (rather than to develop a particular program) is an *experiment*. We will require that an experiment involve an explicit *hypothesis* that makes an assertion about the program development process, just as testing requires a specification.

For example, suppose we were writing sorting programs, that is, programs to satisfy the specification of arranging an array of numbers into ascending order. Running a program to see if it correctly sorts its input would be considered testing. Seeing which of two programs runs faster on a particular set of data would be classified as measuring. We might also formulate the hypothesis that our approach to testing will catch over 90% of all potential programming bugs and then conduct an experiment by running a collection of bug-ridden sorting programs through our testing process and seeing how many are caught.

Three Kinds of Software Mistakes. A program *error* is a specification-program pair with the property that under certain conditions of input and execution environment the program's output does not meet the specification. Thus the term will not be applicable to a program having no specification. (We will also speak of the "error" in a number, meaning the difference between an approximate number and the true value, but it will be easy to distinguish these two uses.) A *mistake* is a program construct that should be changed because it is contrary to the programmer's intent. Thus an error need not be a mistake (since it might only signal that the specification is not appropriate), and a mistake need not be an error (since it might affect efficiency instead of input/output behavior).

We will classify mistakes according to how they are understood. A *blunder* can be explained in terms of an idealized model of computation, e.g., one which assumes that arithmetic is exact and that FORTRAN programs are executed directly by the computer hardware. For instance, we classify the accident of replacing a + by −, or of writing INTGER instead of INTEGER, as a blunder. To understand an *oversight*, on the other hand, requires a grasp of certain details of actual computers, for instance, specifics about arithmetic units or FORTRAN compilers. In particular, a *numerical oversight* is a mistake that hinges on differences between ideal arithmetic and arithmetic as performed by computers. A *portability oversight* is a mistake because of differences, in hardware or software, among computer systems. See Exercises 3 and 4 for examples of portability oversights.

Experiments about Typographical Mistakes. The following changes to a program statement are defined to be *typographical changes:*

1. Replace the value of a binary arithmetic operation by one of the operands
2. Replace one of the arithmetic operators $+$, $-$, $*$, $/$, or $**$ by another
3. Replace one of the relational operators $=$, $\neq$, $<$, $\leq$, $>$, $\geq$ by another
4. Add 1 to a constant or subtract 1 from a constant
5. Replace one occurrence of a scalar (i.e., nonsubscripted variable) by another scalar that appears in the program

For instance, the following statements can be derived from the FORTRAN statement

```
IF (X + Y .GT. 1.0) Z = 0.0
```

by making a typographical change of the indicated type:

Statement	Type of Change
IF (X .GT. 1.0) Z = 0.0	1
IF (X - Y .GT. 1.0) Z = 0.0	2
IF (X + Y .LT. 1.0) Z = 0.0	3
IF (X + Y .GT. 2.0) Z = 0.0	4
IF (X + Y .GT. 1.0) X = 0.0	5

A *mutant* of a given program is another program that can be derived by making a single typographical change to a statement of the original program. A *mutation experiment* is conducted using the following objects:

1. A collection of one or more programs for a certain computational problem
2. A collection of one or more sets of data for that problem
3. Rules for deciding if a computed solution for that problem is *acceptable*

The mutation experiment consists of (1) generating all mutants of the given procedures, (2) executing every mutant on every one of the given sets of data, and (3) determining which (or how many) mutants "survive." A mutant is said to *survive* the experiment if for each of the given sets of data it (a) performs a valid computation (does not divide by zero, generate an out-of-bounds subscript, reference an undefined variable, etc.) and (b) produces an acceptable answer. In brief, the survival of a mutant indicates that the given collection of test data sets is inadequate to distinguish the mutant from the original program.

The software system described in "Mutation Analysis: Ideas, Examples, Problems and Prospects" by T. Budd (in *Software Testing*, edited by B. Chandrasakeran and S. Radicchi, North-Holland, Amsterdam, 1981) almost completely automates the process of conducting mutation experiments on FORTRAN programs. One of the few tasks that must be performed manually is determining which of the surviving mutants are equivalent to the original program, i.e., which of the typographical changes are not mistakes. (For instance, changing an occurrence of $\geq$ to $>$ might not affect the program's quality.) Usually this task is quite simple.

The mutation analysis tool used in experiments described throughout the book was originally developed as an aid for a particular approach to systematic generation of test data in a production environment. The survival of a mutant would be taken as indicating a weakness in the test cases that should be remedied before the program could be considered well tested. Because of the cost (in terms of computer resources) of using this tool, it has not been widely applied in that context. Here we are using the mutation tool as an experimental apparatus to explore both the effectiveness of various test data generation methods and the difficulty of adequately testing various kinds of programs for various kinds of errors. In this laboratory setting the tool has proven to be invaluable.

EXERCISES

1. Consider the following FORTRAN 77 *random number generator*:

```
FUNCTION RAN()
SAVE K
DATA K / 100001 /
K = MOD(K*125,2796203)
RAN = REAL(K) / 2796203.0
RETURN
END
```

[*MOD(K * 125,2796203)* is the remainder after division of *K * 125* by *2796203*.]
(a) Show that any value returned by *RAN* lies between 0 and 1, given certain assumptions about the number of bits in a FORTRAN integer.
(b) Does the statement that *RAN* returns a value between 0 and 1 constitute a specification in the sense discussed in this section? Explain.
(c) Give an informal, yet plausible, condition that the sequence of numbers generated by

```
1   PRINT *, RAN()
    GO TO 1
```

should meet to exhibit "randomness." Execute *RAN* to see whether it meets this condition.
(d) Can you give a specification (that is, a precise and complete statement of requirements) that a random number generator should satisfy? What test would determine whether or not *RAN* is correct in this sense? See *The Art of Computer Programming, Vol. 2: Seminumerical Algorithms*, by Donald E. Knuth (Addison-Wesley, Reading, Mass., 1981) for a discussion of the subtleties of random number generators.
(e) Suppose that a mistake is made when typing *RAN*, causing 100001 to be replaced by 10001. Would that necessarily be considered an error in the sense used here?

2. (a) Let P be a program containing p binary arithmetic operations, q occurrences of relational operators, r occurrences of constants, s occurrences of scalars, and t distinct scalars. Show that P has $6p + 5q + 2r + (t - 1)s$ mutants.
    (b) Show that the following program has 185 mutants.

$$d \leftarrow b^2 - 4c$$
$$\text{if } d \geq 0$$
$$e \leftarrow \sqrt{d}$$
$$r \leftarrow \frac{-b + e}{2}$$
$$s \leftarrow \frac{-b - e}{2}$$
$$\text{else}$$
$$p \leftarrow \frac{-b}{2}$$
$$q \leftarrow \frac{\sqrt{-d}}{2}$$

3. (Optional) The 1966 FORTRAN standard did not include the *SAVE* statement, and FORTRAN compilers differed in the ways they handled *DATA* statements. Specifically, in some environments a variable in a *DATA* statement was reinitialized every time the procedure was entered; in other environments the variable's value at a procedure invocation was its value at termination of the preceding invocation. Discuss the repercussions for the procedure *RAN* of Exercise 1 with the *SAVE* statement removed.

4. (Optional) FORTRAN programs are collections of external subprograms that share data through procedure arguments or *COMMON* (that is, global) data areas. Some FORTRAN compilers handle unsubscripted variables with "call by reference" (i.e., by passing the addresses of procedure arguments), while others use "copy-in, copy-out" (i.e., upon entry to a subprogram, an initial value for the argument is copied into a temporary location to be used throughout execution of the subprogram; upon return, the final contents of the temporary are copied back into the actual argument). This difference among FORTRAN compilers results in two situations where a program's output depends on the compiler's implementation. One case occurs when a variable appears more than once as an actual argument in a procedure reference, as in *CALL THUD(X, X)*. The other case is when an actual argument passed to a subprogram is in a *COMMON* area to which the called subprogram has access. Give programs that illustrate each of these cases. In each case, specify a set of data and the differing outputs.

1.2 AN EXAMPLE OF A TEST AND TWO EXPERIMENTS

In this section we will illustrate the notions of *specification, algorithm, implementation, test, hypothesis, error, numerical oversight, experiment,* and *mutation*

Sec. 1.2 An Example of a Test and Two Experiments

experiment as they arise during development of programs to solve quadratic equations of the form

$$x^2 + bx + c = 0$$

In addition, the following general points about numerical software are illustrated by this example:

1. It may require some effort to determine an appropriate specification.
2. It may be difficult in practice to decide whether the output from a test with a particular set of input data has adhered to a given specification.

A computer can represent numbers to only a limited precision and in general cannot represent the exact solutions of $x^2 + bx + c = 0$ even in cases where b and c are representable. With this limitation in mind, we might propose Specification 1.1a.

Specification 1.1a
The computed solutions of $x^2 + bx + c = 0$ should agree to within the computer's precision with the true solutions.

However, Specification 1.1a is not particularly useful because, as it turns out, no program can satisfy it (unless extra precision is used in the computation). It is more helpful to work with the following specification, which is not the sort of assertion that one is likely to think of right away:

Specification 1.1b
The computed solutions of $x^2 + bx + c = 0$ should agree to within the computer's precision with the true solutions of $x^2 + bx + C = 0$, where C is some number that agrees with c to within the computer's precision.

Contemplate Specification 1.1b awhile and work Exercise 1. We think that you will come to agree with point 1 above.

The mathematical theory that underlies our quadratic equation procedures consists of the following fact:

Theorem 1.1
If b and c are real numbers, then the equation $x^2 + bx + c = 0$ is satisfied by exactly two (possibly identical) numbers, namely

$$\frac{-b \pm e}{2}$$

where $e = \sqrt{d}$ and $d = b^2 - 4c$. In particular, if $b^2 - 4c < 0$, then the two solutions are the complex numbers

$$p \pm i \times q,$$

where $p = -b/2$, $q = \sqrt{-d}/2$, and $i^2 = -1$.

This fact immediately suggests Algorithm 1.1a:

Algorithm 1.1a

$$d \leftarrow b^2 - 4c$$

if $d \geq 0$

$$e \leftarrow \sqrt{d}$$

$$r_1 \leftarrow \frac{-b + e}{2}$$

$$r_2 \leftarrow \frac{-b - e}{2}$$

else

$$p \leftarrow \frac{-b}{2}$$

$$q \leftarrow \frac{\sqrt{-d}}{2}$$

It turns out that Algorithm 1.1a is not as accurate as it should be (see Exercise 1), which illustrates the fact that straightforward embodiment of mathematically correct formulas often results in a poor computer program.

The following variant of Algorithm 1.1a is *correct*, in the sense that it satisfies Specification 1.1b. The modified procedure computes one solution, call it r_1, by a formula that can be guaranteed to be accurate. It follows from basic algebra that the other solution is c/r_1. This approach is followed in Algorithm 1.1b.

Implementing Algorithm 1.1b requires, among other things, that a decision be made about how the calling program is to be notified when the roots are complex. The quadratic equation solver might always return two complex numbers, perhaps having imaginary parts equal to zero, or it might always return two real numbers, p and q, and an indication of whether the roots are p and q or are the complex conjugate pair $p \pm i \times q$.

To *test* whether an implementation of Algorithm 1.1b satisfies Specification 1.1b, we might generate, by some means, a sequence (b_1, c_1), $(b_2, c_2), \ldots, (b_k, c_k)$ of sets of test data. For each set we could apply the quadratic equation solver and check the specification. But how is this check to be made? In other words, given a set b, c of data and computed solutions r_1 and r_2, how can you tell whether Specification 1.1b is satisfied? Not only is it hard to see how this check might be performed in exact arithmetic, but the check is itself a numerical computation that is subject to the same implementation annoyances and susceptibility to arithmetic inaccuracies that plague algorithms. Even designing a test for Specification 1.1a, which is appreciably

Algorithm 1.1b

$d \leftarrow b^2 - 4c$

if $d \geq 0$

    $e \leftarrow \sqrt{d}$

    if $b \geq 0$

        $r_1 \leftarrow \dfrac{-b - e}{2}$

    else

        $r_1 \leftarrow \dfrac{-b + e}{2}$

    $r_2 \leftarrow \dfrac{c}{r_1}$

else

    $p \leftarrow \dfrac{-b}{2}$

    $q \leftarrow \dfrac{\sqrt{-d}}{2}$

easier to check than is Specification 1.1b, requires some thought (see Exercise 2). Here, as is often the case, the intellectual effort required to produce a reliable program to check that the computed solution meets the specification may well equal or exceed that required to understand and implement the algorithm.

Experience indicates that even extensive testing can fail to find program mistakes caused by use of inexact arithmetic. This observation might lead us to formulate the following experimental hypothesis:

Hypothesis 1.1a
Errors resulting from numerical oversights in procedures to solve quadratic equations are hard to uncover through testing.

To conduct an *experiment* to investigate Hypothesis 1.1a we might see how many sets of test data are needed to expose some known *numerical oversights*. For instance, we might use any of the following *errors*:

1. Specification 1.1a and an implementation of Algorithm 1.1a,
2. Specification 1.1a and an implementation of Algorithm 1.1b, or
3. Specification 1.1b and an implementation of Algorithm 1.1a.

(Remember, we are using the term *error* in a special way; an error consists of a specification and a nonconforming program.) Such an experiment differs from testing because, among other things, it requires the use of known program errors, so the specifications being investigated need not be satisfiable and the quadratic equation solvers being executed need not be candidates for general use.

Our experiment might repeat the following process for each of the three errors: Apply the given algorithm to each of 100 "random" sets (b_1, c_1), ..., (b_{100}, c_{100}) of data and check whether the computed results satisfy the given specification. If, say, only 1% of the sets of data reveal the error (in the sense that the corresponding computed solutions do not meet the specification), then the experiment could be interpreted as supporting Hypothesis 1.1a. (As it turns out, the problem of solving a quadratic equation is easy enough that a simple theoretical analysis determines which sets b, c of data can expose these errors. See Example 1 of Section 2.4 for such a theoretical corroboration of Hypothesis 1.1a.)

Many potential programming mistakes in quadratic equation procedures, especially mistakes that are not related to the use of inexact arithmetic, can be detected by even the most naive approach to program development. The following specific claim can be subjected to experimentation.

Hypothesis 1.1b
To expose almost any typographical mistake in a quadratic equation solver, it is sufficient to apply the procedure to one quadratic with real roots and one quadratic with complex roots.

A *mutation experiment* to investigate Hypothesis 1.1b might involve (1) a FORTRAN implementation of Algorithm 1.1a, (2) the quadratics $x^2 - 5x + 6$ and $x^2 + 4x + 5$, and (3) the acceptance criterion that the errors in the two computed values not exceed 0.0001. This experiment will determine that 8 of the 185 typographical changes will not be exposed by either set of data. (See Exercise 2 of the previous section and Exercise 3, following.) At that point one has such options as claiming that the experiment supports Hypothesis 1.1b, reformulating or discarding Hypothesis 1.1b, designing a mutation experiment with a different pair of quadratics (when you solve Exercise 3, you will see why using small integers for test data is not a good idea), etc.

EXERCISES

1. Imagine a hypothetical computer that carries six digits of precision. Thus, if the computer attempts to add 1.23456 and 0.111111, the best it can do is to produce 1.34567 since it cannot represent 1.345671. Numbers are represented in "scientific notation," e.g., $10^6 \times 0.123456$ or $-10^{-3} \times 0.111111$, so the restriction to six digits of precision does not limit the *size* of the numbers that can be computed.

Sec. 1.3   An Example of Performance Measurement   11

(a) What would this computer produce if it applied Algorithm 1.1a to solve $x^2 - 10^4 x + 1 = 0$? What about Algorithm 1.1b? In each case determine how many significant digits of the computed solutions are correct.

(b) Show that on this computer Algorithm 1.1b produces $r_1 = r_2 = 1.66365$ when applied to

$$(x - 1.6625)(x - 1.6648) = x^2 - 3.3273x + 2.76773 = 0$$

(whose exact roots are 1.6625 and 1.6648).

(c) Verify that $(x - 1.66365)^2 = x^2 - 3.3273x + 2.767731\ldots$.

(d) What statements made in this section are illustrated by parts (b) and (c)?

2. (a) Give a plausibility argument showing that if x differs only slightly from y, then the number of significant decimal digits to which x and y agree is indicated by the value of the FORTRAN expression $-LOG10(ABS((X - Y)/X))$. For instance, the formula says that $x = 54.321$ and $y = 54.123$ agree to $2.438\ldots$ decimal places.

(b) Suppose that you are to test whether a single precision implementation of Algorithm 1.1a satisfies Specification 1.1a. Thus you can compute the solutions in double precision to check the accuracy of the single precision solutions. Using part (a), devise a suitable test procedure. Keep in mind that when the solutions of a quadratic equation are complex, their real part $-b/2$ will be computed exactly (on machines using base-2 arithmetic), so that a blind application of this expression to determine the accuracy of the roots' real parts will result in an attempt to compute $LOG10(0.0)$.

(c) Which is harder to understand, Algorithm 1.1a or a program to test adherence to Specification 1.1a?

3. Show that the following typographical changes in Algorithm 1.1 leave it able to correctly solve $x^2 - 5x + 6 = 0$ and $x^2 + 4x + 5 = 0$.

Change *if* $d \geq 0$ to *if* $d > 0$.
Change the 0 in *if* $d \geq 0$ to either 1 or -1.
Change the e in $r_1 \leftarrow (-b + e)/2$ to d.
Change $r_2 \leftarrow (-b - e)/2$ to $r_2 \leftarrow (-b - d)/2$, to $r_2 \leftarrow 2$, or to $r_2 \leftarrow -b - e - 2$.
Change $p \leftarrow -b/2$ to $p \leftarrow -b + 2$.

Why is the first change not a mistake (in the sense that the algorithm's quality is not appreciably affected)?

1.3 AN EXAMPLE OF PERFORMANCE MEASUREMENT

The purpose of this section is to discuss an example of *measurement* and to illustrate the following general points about numerical software:

1. Many, perhaps most, numerical problems are in theory unsolvable. That is, although there may exist computational procedures that often compute acceptable solutions, any procedure to solve the problem would

occasionally fail without giving any indication that failure has occurred. This deficiency has nothing to do with the fact that computer arithmetic is inexact; undiagnosed failures will occur even for computations performed in exact arithmetic.

2. Many useful numerical procedures have the unfortunate property that there exist no useful specifications (in the sense of *precise* and *complete* assertions) describing the procedure's behavior. In brief, we cannot say exactly what the procedure does.

3. When performance measurements are obtained for several programs that purport to solve the same problem, it often happens that no program stands out as better than the others. It may be that each method has a limited domain on which it is substantially better than the other methods.

Our discussion will center on the following computational problem:

The root bracketing problem. Given a function f, find numbers *a* and *b* such that f(*a*) and f(*b*) have opposite signs. If no such numbers exist, then report that no root exists. [A *root* of f is a number *x* such that f(*x*) = 0.]

Unsolvability of the General Problem. Any procedure to solve the root bracketing problem is doomed to fail, at least occasionally. In brief, the problem is that when the program decides that no root exists, the decision must be made without complete knowledge of f and hence can be wrong.

More specifically, here is how to expose the inadequacy of any root bracketing procedure. Apply it to the function f whose value is 1 for all arguments x. If the procedure fails to report that no root exists, then it is incorrect. Otherwise, we can keep a record of all the arguments $x_1, x_2, \ldots, x_k$ where f is sampled. Now rerun the procedure on the function

$$g(x) = 1 - (x - x_1)(x - x_2)\ldots(x - x_k)$$

See Figure 1.1. Since $g(x_1) = g(x_2) = \cdots = g(x_k) = 1$, the procedure will exactly duplicate the earlier computation and report, this time erroneously, that no root exists. (One might argue that a procedure could sample f at every

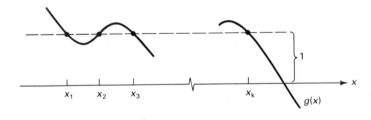

Figure 1.1

machine-representable number x before reporting that no solution exists. However, we are considering only *practical* algorithms. In addition, we are not considering algorithms that generate two different sets of sample arguments when run twice on the same problem. We know of no such "randomized" algorithm that is competitive with "deterministic" algorithms.)

The preceding argument for unsolvability of the root bracketing problem is admittedly informal. However, there are a number of ways of formalizing the notion of "unsolvable problem" and of rigorously proving that the root bracketing problem is unsolvable in that precise sense. Two completely different approaches can be found in the following papers by Webb Miller: "Toward Abstract Numerical Analysis" (*Journal of the Association for Computing Machinery*, July 1973, pp. 399–408) and "Recursive Function Theory and Numerical Analysis" (*Journal of Computer and System Sciences*, Oct. 1970, pp. 465–472).

The Lack of Specifications. Knowing that any procedure to solve the root bracketing problem will sometimes fail, let us turn to the problem of determining the conditions under which a given root bracketing program will work. Here we run up against the difficulty mentioned in point 2; we are unable to make precise and complete statements about the performance of good root bracketing programs.

It is possible to make *vague* claims about a program, such as, the following: *Program X solves the root bracketing problem for most functions f that arise in practice.* Also, precise but narrowly applicable assertions can sometimes be rigorously verified. For instance, it might be possible to prove that if the second derivative of f satisfies $|f''(x)| \leq 10$ for all x, then program X will solve the root bracketing problem for f. However, precise and complete statements about actual behavior are not available.

This disparity between theory and practice is not caused by our inability to prove desired results. Rather, good procedures to solve the root bracketing problem are far more reliable in practice than we know how to *state*. Centuries of work by mathematicians on classifying functions f have not uncovered an appropriate set of concepts.

Indecisive Performance Measurements. The next difficulty arises when we accept the fact that no specification can be found and turn to measuring program performance on specific sets of data. It is often the case that one program is strong in some respects and weak in others and that relative superiority depends on the class of data being used.

For instance, in the course of comparing the performance of two root bracketing programs, call them X and Y, we measured their performance on 100 functions $f_1, f_2, \ldots, f_{100}$, each having a root. (The measuring process we used is described in detail in Programming Assignment 2 of Section 5.4.) A program's *reliability* was indicated by the number of times it successfully solved the root bracketing problem, while its *cost* was measured by the average number of times it sampled f. The measurements came out as follows:

	Program X	Program Y
Reliability	98	100
Cost	38.60	9.73

The statistics for program X show that on two occasions it erroneously reported that no root exists and that it evaluated the functions f_i a total of 3,860 times. The table can be summarized by saying that for the functions $f_1, f_2, \ldots, f_{100}$, program Y was completely reliable, while program X was not, and Y cost only about one fourth as much as X to run.

The measurement process was then repeated using a different set of 100 functions as data, giving the following measurements:

	Program X	Program Y
Reliability	44	15
Cost	50.87	74.89

In these measurements, program X appeared much more reliable than Y and cost less to operate.

Performance measurements of numerical software commonly result in situations of this sort. The more measurements one takes, the murkier the picture becomes. There is no easy answer; hard work is required to make meaningful measurements, and restraint is needed when making claims based on those measurements.

EXERCISE

1. Let P be any root bracketing program. Show that P can be modified to produce a root bracketing program Q with the following properties:
   (a) If f is any function for which P correctly brackets a root, then Q also works correctly for f.
   (b) There exists a function g such that Q brackets a root of g, while P fails to do so.
   In other words, show that for any root bracketing program there is another that is more reliable. [*Hint*: Let P be any root bracketing program and let g be a function that has a root x_0 but for which P is unsuccessful at solving the root bracketing problem. Augment P so that it now finds x_0 when given g.]

1.4 OUR CHOICE OF TOPICS

This book concentrates on the testing and performance measurement of numerical software. In other words, it revolves around the following questions:

Sec. 1.4   Our Choice of Topics

What claims can we make about the software we write? How can we gather evidence to support those claims? Implementation considerations are the other central topic.

Each of Chapters 3-6 begins with one or more straightforward numerical algorithms and then turns to implementation issues. These discussions include the results of mutation experiments designed to provide insight into the process of checking a program for implementation blunders. Generally, the last half of the chapter is devoted to questions about testing or performance measurement.

Chapters 3-6 are organized to proceed, in the following systematic manner, from software whose behavior can be thoroughly analyzed to software whose behavior, to a large extent, must be discovered empirically.

Computation of the Sine Function.   Chapter 3 probes the relationship between the analysis of an algorithm and the testing of a program that implements the algorithm. The chapter illustrates the effort that goes into such analyses and demonstrates the care that may be needed to adequately test a program. You will be amazed by the amount of work required to produce high-quality numerical software.

Linear Equations.   Chapter 4 investigates the reliability of testing as a means for deciding whether a program satisfies a specification; no algorithm analysis is attempted. The chapter also provides the first glimpses of performance measurement.

Solving a Nonlinear Equation.   Chapter 5 forms the bridge between our discussions of testing and of performance measurement. Particular attention is given to general properties of the process of measuring the performance of numerical software.

Automatic Integration.   Chapter 6 takes a harder look at performance measurement. Although no specifications exist for automatic integration procedures, useful claims can be made about the performance of such programs. The chapter considers the nature of those claims and the process of justifying them.

Readers interested in pursuing topics introduced in this book might begin their study with the *ACM (Association for Computing Machinery) Transactions on Mathematical Software*, which has been published quarterly since 1975. In addition, each of the following books contains at least a few papers devoted to these subjects:

Mathematical Software, edited by John R. Rice (Academic Press, New York, 1971)

Software for Numerical Mathematics, edited by D. J. Evans (Academic Press, New York, 1974)

Mathematical Software III, edited by John R. Rice (Academic Press, New York, 1977)

Portability of Numerical Software, edited by Wayne Cowell (Springer-Verlag, Berlin, 1977)

Numerical Software—Needs and Availability, edited by D. A. H. Jacobs (Academic Press, New York, 1978)

Performance Evaluation of Numerical Software, edited by Lloyd Fosdick (North-Holland, Amsterdam, 1979)

Production and Assessment of Numerical Software, edited by M. A. Hennell and L. M. Delves (Academic Press, New York, 1980)

The Relationship Between Numerical Computation and Programming Languages, edited by John Reid (North-Holland, Amsterdam, 1982)

Many prominent issues in modern engineering practice for the production of numerical software are not covered in this book. Here is a list of references for important subjects that we have not covered.

Availability of Packaged Software. *Sources and Development of Mathematical Software*, edited by Wayne Cowell (Prentice-Hall, Englewood Cliffs, N.J., 1984) is a gold mine of information about numerical software.

Classical Numerical Analysis. *Numerical Methods* by Germund Dahlquist, Ake Björck, and Ned Anderson (Prentice-Hall, Englewood Cliffs, N.J., 1974) is one of many good books in this area.

Portability Oversights. "FORTRAN 77 Portability" by J. Larmouth (*Software—Practice and Experience*, Oct. 1981, pp. 1071–1117).

Program Documentation. "Solving Nonstiff Ordinary Differential Equations—the State of the Art" by L. F. Shampine, H. A. Watts, and S. M. Davenport (*SIAM Review*, July 1976, pp. 376–411), an excellent comparison of numerical software, treats a number of issues besides program documentation.

Programming Style. *The Elements of Programming Style* by Brian Kernighan and P. J. Plauger (McGraw-Hill, New York, 1978) or *The Elements of FORTRAN Style* by Charles Kreitzberg and Ben Shneiderman (Harcourt Brace Jovanovich, New York, 1972).

Traditional Software Engineering. See the following two papers by William Howden: "Applicability of Software Validation Techniques to Scientific Programs" (*ACM Transactions on Programming Languages and Systems*, July 1980, pp. 307–320) and "Validation of Scientific Programs" (*ACM Computing Surveys*, June 1982, pp. 193–227).

2
FLOATING-POINT ARITHMETIC

Approaches for dealing with limited-precision arithmetic are gathered together in this chapter. Section 2.1 presents our idealized model of the way that the computer represents and operates on real numbers. In practice, few computers satisfy all these assumptions, but most come close enough that the conclusions drawn throughout the book accurately predict actual program performance. Reasoning about cleanly designed arithmetic units is hard enough without taking into account such anomalies as $1.0 \times x \neq x$ or $x \times y \neq y \times x$ (which actually happen on some computers). However, remember *Murphy's law of floating-point arithmetic*: Anything that can go wrong, does on some computer. Our prescription for diseased arithmetic facilities is to rigorously test the validity of conclusions about numerical accuracy.

Section 2.2 presents a collection of implementation techniques aimed at simplifying the task of moving numerical software among computers having different floating-point number systems. These techniques allow many numerical programs to be completely portable, in the sense of running correctly on a wide variety of computers. Moreover, extensive libraries of several hundred numerical procedures can be written so that adaptation to a particular machine requires only a handful of text editor instructions. One of the uses that we will make of such techniques is to write portable *test harnesses* (i.e., programs whose purpose is to test applications software such as sine procedures, linear equation solvers, etc.).

Section 2.3 discusses a way of computing how many floating-point numbers lie between given numbers x and y. This method allows us to assess the accuracy of x as an approximation to y relative to the computer's precision.

Section 2.4 introduces a simple approach to error analysis that explains the behavior of both successful and unsuccessful sine procedures and allows us to verify the reliability of several test harnesses. On the other hand, examples are given for which this way of thinking about error propagation is woefully inadequate. In such cases, the choice is between more difficult theoretical techniques and experimentation.

2.1 OUR ASSUMPTIONS ABOUT FLOATING-POINT ARITHMETIC

We will think of a floating-point number system as characterized by four parameters, namely the *base* (or *radix*) β, the *precision p*, the *minimum exponent m*, and the *maximum exponent M*. The set of floating-point numbers determined by β, p, m, and M consists of the number zero and all numbers x having a base-β representation of the form

$$x = \pm \beta^e \times 0.d_1 d_2 \cdots d_p$$

where $d_1, d_2, \ldots, d_p$ are base-β digits (namely, $0, 1, 2, \ldots, \beta - 1$), $d_1 \neq 0$, and the *exponent e* satisfies $m \leq e \leq M$. We will call $0.d_1 d_2 \cdots d_p$ the *floating-point fraction* of x; it represents the number

$$d = d_1 \beta^{-1} + d_2 \beta^{-2} + \cdots + d_p \beta^{-p}$$

which satisfies $\beta^{-1} \leq d < 1$.

For example, consider the floating-point number system determined by $\beta = 2$, $p = 3$, $m = -1$, and $M = 1$. This system contains zero and all numbers with binary representation

$$x = \pm 2^e \times 0.1 d_2 d_3$$

where $-1 \leq e \leq 1$ and d_2 and d_3 are each 0 or 1. Thus for nonzero x there are two choices for the sign ($+$ or $-$), three choices for the exponent e ($-1, 0$, or 1), and four choices for the floating-point fraction (the base-2 fractions 0.100, 0.101, 0.110, and 0.111), giving $1 + 2 \times 3 \times 4 = 25$ floating-point numbers. Converting these fractions to their decimal ($\beta = 10$) representations makes them easier to understand, without changing their numerical values. In decimal, the four permissible floating-point fractions are $\frac{1}{2}, \frac{5}{8}, \frac{3}{4}$, and $\frac{7}{8}$. For example, the binary fraction 0.101 represents $1 \times 2^{-1} + 0 \times 2^{-2} + 1 \times 2^{-3} = \frac{1}{2} + \frac{1}{8} = \frac{5}{8}$. Thus the smallest positive floating-point number is $\sigma = 2^{-1} \times \frac{1}{2} = \frac{1}{4}$, while the largest is $\lambda = 2^1 \times \frac{7}{8} = 1\frac{3}{4}$. The nonnegative floating-point numbers can be depicted as shown in Figure 2.1. This picture is misleading in that the distance ϵ from 1 to the next larger floating-point number equals σ.

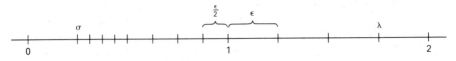

Figure 2.1

Sec. 2.1 Our Assumptions about Floating-Point Arithmetic

Moreover, the picture does not highlight the unique character of the spacing between 0 and its neighbors. In a floating-point number system with $\beta = 10$, $p = 6$, and $m = -100$, which gives a much better feel for realistic floating-point systems, we would have $\epsilon = 10^{-5}$ and $\sigma = 10^{-101}$. Moreover, there would be no floating-point numbers between 0 and σ but 899,999 of them between σ and 10σ.

The distance ϵ from 1 to the next larger floating-point number is called *machine epsilon*. It is probably the most useful quantity associated with a floating-point number system because it gives a measurement of the "granularity" of the floating-point system that is valid over the entire range of nonzero machine numbers.

The Most Important Fact About Floating-Point Number Systems
The spacing between a floating-point number x and an adjacent floating-point number is at least $\epsilon|x|/\beta$ and at most $\epsilon|x|$ (unless x or the neighbor is 0).

Thus the spacing ϵ/β to the left of 1.0 and the spacing ϵ to the right of 1.0 are the extreme cases of relative spacing between consecutive nonzero floating-point numbers.

We will make the following assumptions about the floating-point operations $+$, $-$, $\times$, $/$, and $\sqrt{}$. Here σ denotes the smallest positive floating-point number, λ denotes the largest floating-point number, and x denotes the true result of an arithmetic operation on given floating-point operands (one operand in the case of $\sqrt{}$).

1. If $\sigma \leq |x| \leq \lambda$, then the computed result of the operation is the floating-point number nearest to x. (If x is midway between two adjacent floating-point numbers, then the computed result can be either neighbor.)
2. If $|x| < \sigma$, then the computed result of the operation is zero ("silent and nondestructive underflow").
3. If $|x| > \lambda$, then the computation is immediately terminated ("fatal overflow").

Any attempt to divide by zero or to take the square root of a negative number is assumed to stop program execution.

EXERCISES*

1. Show that if $m \leq e < M$, then there are exactly $(\beta - 1)\beta^{p-1} + 1$ floating-point numbers x satisfying $\beta^{e-1} \leq x \leq \beta^e$ and that these numbers have equal spacing β^{e-p}. As a special case, show that $\epsilon = \beta^{1-p}$.
2. Prove the most important fact about floating-point number systems.
3. Show that $\sigma = \beta^{m-1}$ and $\lambda = \beta^M(1 - \beta^{-p})$.

*The reader who skips Exercise 1 or 2 will soon become lost.

4. (Optional) Assume that $\beta = 2$, $p = 24$, and $m = -128$. The *bisection process* goes as follows.

Initialize a and b to certain values.

repeat

$$c \leftarrow \frac{a+b}{2}$$

either $a \leftarrow c$ or $b \leftarrow c$

   (a) Show that the bisection process can be repeated 23 times beginning with $a = 1$ and $b = 2$ before adjacent machine numbers are obtained. On the other hand, show that only 13 bisection steps will be taken if initially $a = 2000$ and $b = 2001$.
   (b) Suppose that the bisection process is applied, beginning with $a = 0$ and $b = 1$. Show that at least 24 bisection steps will be taken before adjacent machine numbers are obtained. Show that 152 is the maximum number of bisection steps that may be taken before adjacent machine numbers are obtained.

2.2 PARAMETERIZING NUMERICAL SOFTWARE

Software portability is facilitated by the practice of writing programs in which as much as possible of the machine dependency is isolated to a few variables, called *environmental parameters*. For example, the FORTRAN 66 command to read from the standard input device is

    READ(100,...

if the program is to be run on a Cray 1, but

    READ(5,...

for IBM machines. If the program instead uses

    READ(INPUT,...

where *INPUT* is a global variable, then a change to one line of the program is sufficient to adapt all *READ*s to a new computer.

In this section we will discuss several ways of making the values of environmental parameters available to a program. Our emphasis will be on currently practical methods for dealing with the environmental parameters of particular concern to the producers of numerical software, though we will briefly mention two approaches that we expect to eventually supplant current practices.

2.2.1 An Example

The environmental parameters β, p, m, M, σ, λ, and especially ϵ are crucial for numerical software. For example, consider the problem of designing a procedure $root(f, a, b, tol)$ that is to compute, to within tol, a solution of $f(x) = 0$ lying between a and b; the program is allowed to assume that $a < b$ and that $f(a)$ and $f(b)$ have opposite sign. The difficulty is that if the user provides a value of tol that is too small, then the procedure may well never terminate. The following procedure, a variant of the *bisection method* of Section 5.1, solves the problem by resetting tol, if necessary, so as to make it at least as large as the spacing between floating-point numbers near the solution:

$\epsilon \leftarrow$ machine epsilon

$mytol \leftarrow max(tol, \epsilon \times |a|, \epsilon \times |b|)$

repeat

$$c \leftarrow \frac{a+b}{2}$$

if $f(a)$ and $f(c)$ have the same sign

$a \leftarrow c$

else

$b \leftarrow c$

until $b - a < 2 \times mytol$

return $\frac{a+b}{2}$

2.2.2 Computing Arithmetic Parameters

Programs can compute the values (for the machine on which the program is being run) of certain environmental parameters. Here we will consider the computation of the environmental parameters used in this book. A program to compute values of nine additional environmental parameters can be found on pp. 258–264 of the book *Software Manual for the Elementary Functions* by William J. Cody, Jr. and William Waite (Prentice-Hall, Englewood Cliffs, N.J., 1980).

The Base and Precision. The program segment given below is adapted from "Algorithms to Reveal Properties of Floating-Point Arithmetic" by Michael Malcolm (*Communications of ACM*, Nov. 1972, pp. 949–951). It uses the fact that the integers that can be represented as floating-point numbers are

$$1, 2, \ldots, \beta^p - 1, \beta^p, \beta^p + \beta, \beta^p + 2\beta, \ldots, \beta^{p+1}, \beta^{p+1} + \beta^2, \ldots$$

For example, on a three-digit decimal system ($p = 3, \beta = 10$) the representable integers are $1, 2, \ldots, 999, 1000, 1010, 1020, \ldots, 10,000, 10,100, \ldots$. For an arbitrary floating-point base β, this fact becomes apparent once the following points are understood:

1. The base-β representation of β^p is $10 \cdots 0$, with p zeros. Thus β^p can be represented as a floating-point number with a fraction of length 1, i.e., as $\beta^{p+1} \times 0.1$.
2. The base-β representation of an integer less than β^p contains p or fewer digits, so the integer can be represented in floating-point format with a fraction of p or fewer digits. For instance, the base-β representation of the integer $\beta^p - 1$ is $dd \cdots d$, with p copies of the digit $d = \beta - 1$. In floating-point format this becomes $\beta^p \times .dd \cdots d$.
3. The integers between β^p and β^{p+1} (but not including β^{p+1}) are just the integers whose base-β representation requires $p + 1$ digits. The integers in that range whose base-β representation ends in a zero are just those that can be represented as base-β floating-point numbers with a fraction of p or fewer digits. For instance, the next floating-point number after β^p is $\beta^{p+1} \times 0.10 \cdots 01$ (with $p - 2$ zeros) $= 10 \cdots 010$, i.e., $\beta^p + \beta$.

The following program segment keeps doubling A until it reaches a point where the spacing between consecutive floating-point numbers exceeds 1. It then looks for the next larger floating-point number, from which it subtracts A to get β. At that point p can be determined as the smallest power of β for which the distance to the next floating-point number exceeds 1.

```
      INTEGER BETA, P
      A = 1.0
10    A = 2.0*A
      IF ((A+1.0)-A .EQ. 1.0) GO TO 10
      B = 1.0
20    B = 2.0*B
      IF (A+B .EQ. A) GO TO 20
      BETA = (A+B) - A
      P = 0
      A = 1.0
30    P = P + 1
      A = A*BETA
      IF ((A+1.0)-A .EQ. 1.0) GO TO 30
```

Suppose that this program were run on a hypothetical computer with three-digit decimal arithmetic. On the tenth iteration of the loop containing statement 10, the exact value 1024 of A would be computed as 1020, and the test at the bottom of the loop would fail. For $B = 2$ or 4, $A + B$ would be computed as 1020, but for $B = 8$, $A + B$ would be computed as 1030. Subtraction would give BETA $= (A + B) - A = 10$, and the third iteration of the final loop would determine that $p = 3$.

Machine Epsilon. If β and p are known, then ϵ can be determined from the formula $\epsilon = \beta^{1-p}$. Alternatively, ϵ can be computed directly by the following FORTRAN program segment:

```
        TRY = 1.0
10      EPS = TRY
        TRY = 0.5*EPS
        TRYP1 = TRY + 1.0
        IF (TRYP1 .GT. 1.0) GO TO 10
        EPSP1 = EPS + 1.0
        EPS = EPSP1 - 1.0
```

On our hypothetical three-digit decimal computer the test $TRY .GT. 1.0$ fails when $TRY = 0.0039$. At that point $EPS = 0.0078$, so the last two statements compute $EPSP1 = 1.01$ and $EPS = 0.01$.

The Integer Overflow Level. Whereas the integers $1, 2, \ldots, \beta^p$ can all be represented as floating-point numbers, in many cases they cannot be represented *as machine integers* (e.g., in the FORTRAN *INTEGER* type). In particular, for double precision floating-point systems, β^p almost always lies far above the integer overflow level. If each integer location is allotted i bits (including one bit for the sign), then the integer overflow level is 2^{i-1}; in other words, the largest permissible integer is $2^{i-1} - 1$.

On many machines, overflow during an integer multiplication evokes no messages: The left-most bits of the exact product are silently dropped. For those machines, the following program segment determines a value that is half of the integer overflow level:

```
        INTEGER HALF,WHOLE
        WHOLE = 1
10      HALF = WHOLE
        WHOLE = 2*HALF
        IF (WHOLE .GT. HALF) GO TO 10
```

Advantages and Disadvantages. When a program is written so that its only machine dependencies occur as environmental parameters whose values are computed, then the program is completely portable in the sense that it will run without change on a new machine. Of course, there are limitations to this approach because the values of some environmental parameters, e.g., the standard input unit, cannot be determined by the program and because some machine dependencies cannot be encapsulated as environmental parameters.

Moreover, some computer-compiler combinations make it difficult to compute the correct values of environmental parameters. For example, on our DEC PDP-11/70 running version 7 of the UNIX* operating system, the value of p computed as above is 56, instead of the correct value 24. This happens because the expression $(A + 1.0) - A$ is evaluated in a double precision register, causing the computed p to be the value appropriate for double precision. If the statement

```
IF ((A+1.0)-A .EQ. 1.0) GO TO 30
```

is changed to

```
AP1 = A + 1.0
IF (AP1-A .EQ. 1.0) GO TO 30
```

then the computed value of $A + 1.0$ is converted to single precision for storage in memory, and $p = 24$ is correctly determined. However, this approach fails if the program is compiled with the "optimization" option (Murphy's law of floating-point arithmetic). A trick that works even with compiler optimization is to use the statement

```
IF (COPY(A+1.0)-A .EQ. 1.0) GO TO 30
```

where *COPY* is defined by

```
FUNCTION COPY(X)
COPY = X
RETURN
END
```

For a discussion of problems of this sort that arise when computing values of environmental parameters, see "More on Algorithms that Reveal Properties of Floating Point Arithmetic Units" by W. Morven Gentleman and Scott Marovich (*Communications of ACM*, May 1974, pp. 276–277).

2.2.3 The PORT Library Approach

The PORT Library, a collection of numerical procedures developed at Bell Laboratories, achieves portability by using two techniques that have been

---

*UNIX is a trademark of Bell Telephone Laboratories.

widely adopted by other producers of numerical software:

1. All PORT programs are sent through a program that checks whether they are written in a particular portable subset of FORTRAN 66 described in "The PFORT Verifier" by Barbara Ryder (*Software—Practice and Experience*, Oct.–Dec. 1974, pp. 359–377). In addition to checking for violations of the 1966 FORTRAN standard and for standard-conforming constructs not supported by some FORTRAN compilers, the PFORT Verifier program diagnoses such portability oversights as those in Exercise 4 of Section 1.1.

2. Machine dependencies in PORT programs are restricted to environmental parameters that are evaluated by function invocations. Integer parameters are delivered to PORT Library programs by a function called *I1MACH*, which provides 16 constants for each of 11 computer families. Another function provides five single precision environmental parameters and a third supplies five double precision values. All three functions are part of a publicly available software package described in "The PORT Mathematical Subroutine Library" by P. A. Fox, A. D. Hall, and N. L. Schryer (*ACM Transactions on Mathematical Software*, June 1978, pp. 104–126).

In particular, *I1MACH(1)* is the standard input unit, so a PORT Library program uses the command

```
    READ (I1MACH(1),...
```

to read a line from the standard input device. The following is excerpted from *I1MACH* with some poetic license (such as using FORTRAN 77 block *IF*s for readability):

```
      INTEGER FUNCTION I1MACH(I)
C
C I/O UNIT NUMBERS.
C
C    I1MACH( 1)=THE STANDARD INPUT UNIT.
C
         ...
C FLOATING-POINT NUMBERS.
C
C    I1MACH(10)=BETA, THE BASE.
C
C    I1MACH(11)=P, THE NUMBER OF BASE-BETA DIGITS.
C
```

```
C     I1MACH(12) = EMIN, THE SMALLEST EXPONENT E.
C
      INTEGER IMACH(16)
         ...
C
C     MACHINE CONSTANTS FOR THE CRAY 1.
C
C     DATA IMACH( 1) /    100 /
         ...
C     DATA IMACH(10) /      2 /
C     DATA IMACH(11) /     48 /
C     DATA IMACH(12) /  -8192 /
         ...
C
C     MACHINE CONSTANTS FOR THE IBM 360 / 370 SERIES,
C     THE XEROX SIGMA 5 / 7 / 9 AND THE SEL SYSTEMS 85 / 86.
C
C     DATA IMACH( 1) /      5 /
         ...
C     DATA IMACH(10) /     16 /
C     DATA IMACH(11) /      6 /
C     DATA IMACH(12) /    -64 /
         ...
C
      IF (I.GT.1 .AND. I.LT.16) THEN
         I1MACH = IMACH(I)
         RETURN
      ELSE
         PRINT *, 'ERROR IN I1MACH -- I OUT OF BOUNDS'
         STOP
      ENDIF
      END
```

To prepare the PORT Library for installation on, say, a Cray 1 requires only that the *C*s at the beginnings of 26 lines be removed (16 in *I1MACH*, 5 each in its real and double precision counterparts), thereby changing the lines from comments to active *DATA* statements.

Advantages. The fact that the PORT Library is running on a wide variety of computers proves the feasibility of the library's approach to portability. Software producers who employ this technique enjoy the advantage of not having to customize software. In other words, all customers receive identical magnetic tapes. Furthermore, the installer's task is trivial if the machine's characteristics are known to PORT; for other machines, requirements for installation are precisely defined.

Sec. 2.2   Parameterizing Numerical Software   27

There is another, rather subtle but quite important, advantage of the particular approach taken by the PORT Library. If the environmental parameters are defined with care, then it is possible for programs written with a clean model of floating-point arithmetic (e.g., the model discussed in the previous section) to work correctly on machines that violate the assumptions of the model. For instance, suppose that a computer represents numbers with a 24-bit fraction, i.e., $\beta = 2$ and $p = 24$, but that instead of exact results being rounded to the closest representable number, excess digits are chopped. The errors in individual operations are then no larger than for a machine with $p = 23$ that employs rounding, so a program supplied with the value $p = 23$ would perform properly. The paper "A Simple but Realistic Model of Floating-Point Computation" by W. S. Brown (*ACM Transactions on Mathematical Software*, Dec. 1981, pp. 445–480) discusses in detail the PORT approach to "penalizing" environmental parameters to compensate for violations of certain axioms about the computer's arithmetic. The problem of finding appropriate settings for these parameters is addressed in "Determination of Correct Floating-Point Model Parameters" by N. L. Schryer (in *Sources and Development of Mathematical Software*, edited by Wayne Cowell, Prentice-Hall, Englewood Cliffs, N.J., 1984, pp. 360–366).

2.2.4 Macro Processors

The task of replacing all occurrences of the character string "INPUT" by "100" (or "5" if desired), changing "BETA" to "2", etc., can be accomplished by a program called a *macro processor*. For instance, upon reading the line

    READ(INPUT, 70) X

the macro processor would print the line

    READ(100, 70) X

whereas a line containing no environmental parameters or other special character strings would be copied verbatim. Machine-specific programs produced this way run faster than if they evaluated environmental parameters by computing them or by invoking functions, though the gain is probably insignificant.

The main advantage of this approach is the generality of macro processors. Arbitrary kinds of program variants (for instance, single and double precision versions of a program) can be generated automatically, as will be illustrated in Section 4.2. Additional examples of the use of a macro processor for numerical software development are given in "Tools for Numerical Programming" by Webb Miller (in *The Relationship Between Numerical Computation and Programming Languages*, edited by John Reid, North-Holland, Amsterdam, 1982, pp. 347–359).

A further benefit from the flexibility of macro processors is that character files other than programs can be transformed. For instance, program documentation can be automatically tailored to a specific machine.

2.2.5 Programming Language Standards

For some time, people have discussed the idea that "environmental inquiries," similar in spirit to the PORT functions described above, should be made part of programming languages, just as the elementary functions $[sin(x), $ etc.] are today. The status in 1981 of progress toward this goal is outlined in "Floating-Point Parameters, Models and Standards" by W. J. Cody (in *The Relationship Between Numerical Computation and Programming Languages*, edited by John Reid, North-Holland, Amsterdam, 1982, pp. 51–65).

2.2.6 Hardware Standardization

The simplest and best, though hardest to attain, solution to the problem of environmental parameters is to standardize floating-point hardware, so that the values of the parameters become universal constants. Surprisingly, real progress is being made along these lines. See the March 1981 issue of *Computer* for discussions about a forthcoming standard for floating-point arithmetic whose details are superior to anything that has gone before.

PROGRAMMING ASSIGNMENT
(Optional and Difficult)

The goal of this assignment is to produce a certain program that involves environmental parameters and that can be used during development of numerical software. One lesson to be learned from this assignment is that it is insufficient to merely provide numerical values for environmental parameters; close attention must be paid to specifying the exact meaning, or permissible use, of those values. This program lies somewhere near the boundary between programs for which attempts to achieve portability are worthwhile and those for which portability is too expensive, or impossible, to attain. We leave it to the reader to form an opinion as to the program's location with respect to that boundary.

The book *Computer Methods for Mathematical Computations* by George Forsythe, Michael Malcolm, and Cleve Moler (Prentice-Hall, Englewood Cliffs, N.J., 1977) contains the following random number generator:

```
      REAL FUNCTION URAND(IY)
      INTEGER IY
C
C     URAND IS A UNIFORM RANDOM NUMBER GENERATOR BASED ON
C THEORY AND SUGGESTIONS GIVEN IN D.E. KNUTH (1969), VOL 2.
```

Sec. 2.2   Parameterizing Numerical Software   29

```
C THE INTEGER IY SHOULD BE INITIALIZED TO AN ARBITRARY
C INTEGER PRIOR TO THE FIRST CALL TO URAND. THE CALLING
C PROGRAM SHOULD NOT ALTER THE VALUE OF IY BETWEEN
C SUBSEQUENT CALLS TO URAND. VALUES OF URAND WILL
C BE RETURNED IN THE INTERVAL (0,1).
C
      INTEGER IA, IC, ITWO, M2, M, MIC
      DOUBLE PRECISION HALFM
      REAL S
      DOUBLE PRECISION DATAN, DSQRT
      DATA M2/0/, ITWO/2/
      IF (M2 .NE. 0) GO TO 20
C
C IF FIRST ENTRY, COMPUTE MACHINE INTEGER WORD LENGTH
C
      M = 1
10    M2 = M
      M = ITWO*M2
      IF (M .GT. M2) GO TO 10
      HALFM = M2
C
C COMPUTE MULTIPLIER AND INCREMENT FOR LINEAR CONGRUENTIAL
C METHOD
C
      IA = 8*IDINT(HALFM*DATAN(1.D0)/8.D0)+5
      IC = 2*IDINT(HALFM*(0.5D0 - DSQRT(3.D0)/6.D0))+1
      MIC = (M2 - IC) + M2
C
C S IS THE SCALE FACTOR FOR CONVERTING TO FLOATING POINT
C
      S = 0.5 / HALFM
C
C COMPUTE THE NEXT RANDOM NUMBER
C
20    IY = IY*IA
C
C THE FOLLOWING STATEMENT IS FOR COMPUTERS WHICH DO NOT ALLOW
C INTEGER OVERFLOW ON ADDITION
C
      IF (IY .GT. MIC) IY = (IY - M2) - M2
C
      IY = IY + IC
C
C THE FOLLOWING STATEMENT IS FOR COMPUTERS WHERE THE WORD
C LENGTH FOR ADDITION IS GREATER THAN FOR MULTIPLICATION
C
```

```
      IF (IY/2 .GT. M2) IY=(IY-M2)-M2
C
C THE FOLLOWING STATEMENT IS FOR COMPUTERS WHERE INTEGER
C OVERFLOW AFFECTS THE SIGN BIT
C
      IF (IY .LT. 0) IY=(IY+M2)+M2
      URAND = FLOAT(IY)*S
      RETURN
      END
```

As was the case with the function *RAN* in Exercise 1 of Section 1.1, *URAND* generates random numbers with the recurrence

$$y \leftarrow a \times y + c \ (modulo\ m)$$

i.e., y is set to the remainder upon division of $a \times y + c$ by m. In *URAND*, m is the integer overflow threshold (that is, it is one larger than the largest representable integer); a has the properties:

1. $a\ (modulo\ 8) = 5$,
2. $a \approx m/2$,
3. The bit pattern of a is the same as that of π;

and c is an odd integer with

$$\frac{c}{m} \approx \frac{1}{2} - \frac{1}{6}\sqrt{3} \approx 0.21132$$

This assignment has the following three parts:

(a) *URAND* depends on certain assumptions about how the FORTRAN system responds to integer overflow. Exactly what are those assumptions? Explain the purpose of each of the three statements that begin "*IF (IY* ···." In particular, should the test *IF (IY .GT. MIC)* read *IF (IY .GE. MIC)*? Also, can the test *IF (IY/2 .GT. M2)* ever return *true*? [Hint: CDC 6000/7000 series computers have hardware for adding 60-bit words representing integers, but form products of integers using hardware for multiplying floating-point numbers, where $\beta = 2$ and $p = 48$.]

(b) Modify *URAND* so that $a \times y + c\ (modulo\ m)$ is computed in such a way that integer overflow is impossible. [*Hint*: Let m equal 2^b. Divide each of a and y into two $b/2$-bit parts. The right-most b bits of $a \times y$ can be assembled out of the values resulting from three multiplications, each involving a pair of $b/2$-bit operands. (A little care is required if b is odd.) This approach is just a special case of "multiple precision arithmetic" as discussed on pp. 229–248 of *The Art of Computer Programming, Vol. 2: Seminumerical Algorithms*, by Donald Knuth (Addison-Wesley, Reading, Mass., 1981).]

(c) Modify your version of *URAND* so that all necessary information about the host computer is derived from the value returned by the following PORT-like function:

```
      INTEGER FUNCTION MACHIN()
C
C INTEGER OPERATIONS ARE EXACT (EXCEPT THAT DIVISION CHOPS
C THE ANSWER TO AN INTEGER) PROVIDED THAT THE OPERANDS AND
C RESULT ARE LESS THAN 2**MACHIN() IN ABSOLUTE VALUE.
C IN OTHER WORDS, AS LONG AS ALL VALUES ARE MACHIN()-BIT
C NUMBERS (NOT COUNTING THEIR SIGNS), ARITHMETIC WORKS
C AS EXPECTED.
C
C FOR INSTALLATION, REMOVE THE APPROPRIATE 'C' FROM COLUMN 1.
C
C FOR THE BURROUGHS 1700 SYSTEM.
C     MACHIN = 33
C
C FOR THE BURROUGHS 5700 / 6700 / 7700 SYSTEMS.
C     MACHIN = 39
C
C FOR THE CDC 6000 / 7000 SERIES.
C     MACHIN = 48
C
C FOR THE CRAY 1.
C     MACHIN = 63
C
C FOR THE DATA GENERAL ECLIPSE S / 200 AND
C FOR PDP-11 FORTRAN'S SUPPORTING 16-BIT INTEGER ARITHMETIC.
C     MACHIN = 15
C
C FOR THE HARRIS 220.
C     MACHIN = 23
C
C FOR THE HONEYWELL 600 / 6000 SERIES, PDP-10, AND
C UNIVAC 1100 SERIES.
C     MACHIN = 35
C
C FOR THE IBM 360 / 370 SERIES, XEROX SIGMA 5 / 7 / 9, SEL
C SYSTEMS 85 / 86 AND FOR PDP-11 FORTRAN'S SUPPORTING
C 32-BIT INTEGER ARITHMETIC.
C     MACHIN = 31
C
      RETURN
      END
```

EXERCISES

1. Do the program segments to compute β, p, and ϵ work correctly on a machine whose only departure from our model of arithmetic is that floating-point fractions of exact results are chopped (i.e., truncated instead of rounded)? Trace the values of program variables when $\beta = 10$ and $p = 3$.
2. In the program segment that computes β, trace the evolution of values of the variables A and B when $\beta = 3$ and $p = 3$. In particular, show that when $BETA$ is computed, $A = 33$ and $B = 2$.
3. Show how to write a FORTRAN subprogram so that it computes machine epsilon just once, the first time it is called.
4. Give a program segment that computes the underflow threshold σ.
5. (Optional) In the program segment that computes β, suppose that the statement $A = 2.0 * A$ (i.e., the statement labeled 10) is replaced by $A = 3.0 * A$.
   (a) Show that if the revised code is run on a machine with $\beta = 2$ and $p = 9$, then values $BETA = 2$ and $P = 9$ will be computed.
   (b) Show that if the revised code is run on a machine with $\beta = 2$ and $p = 10$, then values $BETA = 4$ and $P = 5$ will be computed.
   (c) Show that the revised code will correctly compute β and p if either (1) $\beta \geq 3$ or (2) $\beta = 2$ and p has the property that some power of 3 falls between β^p and β^{p+1}.

2.3 ULPS

In this section we will discuss the notion of two numbers differing by k *units in the last place*, written "k ulps," relative to some floating-point number system.

The Definition of ulps(x, y). Although the intuitive idea of "units in the last place" is clear in most cases, it is ambiguous in others. For instance, consider a system of three-digit decimal numbers (that is, $p = 3$, $\beta = 10$). The floating-point numbers around 1 are

$$\ldots, 0.995, 0.996, 0.997, 0.998, 0.999, 1.00, 1.01, 1.02, \ldots$$

Clearly 0.996 differs from 0.997 by one ulp and from 0.999 by three ulps. Our intuition breaks down when comparing, say, 0.998 and 1.01. Do they differ by 12 ulps (since one ulp of 0.998 equals 0.001 and $1.01 = 0.998 + 12 \times 0.001$)? Do they differ by 1.2 ulps (since one ulp of 1.01 equals 0.01 and $0.998 = 1.01 - 1.2 \times 0.01$)? Do they differ by three ulps (since there are exactly two floating-point numbers between them)?

All potential ambiguity in the intuitive notion of ulps is removed by the definition in the following paragraph. It gives a precise meaning to the assertion "x and y differ by k ulps" even when x and y are not floating-point numbers. This extra flexibility is useful when, for example, we are using a "true solution" computed in double precision to measure the accuracy of a

Sec. 2.3   Ulps   33

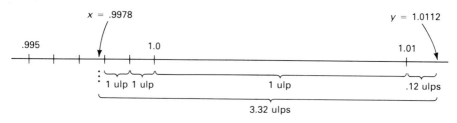

Figure 2.2

solution computed in single precision. The error is measured in "single precision ulps," and the double precision value is almost certainly not a single precision number. (For example, if $\beta = 2$, single precision $p = 24$, and double precision $p = 54$, then there are about a billion double precision numbers between a pair of consecutive single precision numbers.)

The members of any fixed set of floating-point numbers divide the real line into intervals. Two of these intervals are infinite, while the others have adjacent machine numbers as end points. Suppose that x and y are numbers having the same sign, and that they lie in the range of machine numbers (that is, their absolute values fall between σ and λ). We will say that x and y differ by k ulps, or $k = ulps(x, y)$, if there are k intervals, including fractions of the interval containing x and the interval containing y, between x and y. If numbers x and y do not lie on the same side of zero, or if at least one of them falls outside the range of machine numbers, then we will say that x and y differ by ∞ ulps. In particular, zero differs from any nonzero number by ∞ ulps.

For example, in the case of three-digit decimal machine numbers with $x = 0.9978$ and $y = 1.0112$, the situation can be pictured as shown in Figure 2.2. In this figure x differs from 0.998 by 0.2 ulp (because the distance from 0.9978 to 0.998 is 0.2 of the length of the interval [0.997, 0.998]), and 1.01 differs from 1.0112 by 0.12 ulp (because 1.0112 is 12% of the way from 1.01 to 1.02), so x differs from y by 3.32 ulps.

Computing ulps(x, y). The following FORTRAN function computes $ulps(x, y)$:

```
      REAL FUNCTION ULPS(X,Y)
      DOUBLE PRECISION X, Y
* COMPUTES THE DIFFERENCE, IN SINGLE PRECISION ULPS,
* BETWEEN DOUBLE PRECISION NUMBERS X AND Y. IF X
* AND Y DIFFER BY MORE THAN A FACTOR OF TWO,
* THEN THE RESULT IS ONLY APPROXIMATE.
      DOUBLE PRECISION ABSX, ABSY
      REAL BSUPE, BSUPE1, EBETA, EPS, SPACE
      INTEGER BETA
```

```
* GET THE VALUES OF ENVIRONMENTAL PARAMETERS FOR SINGLE
* PRECISION.
      BETA = ?
      EPS = ?
      EBETA = BETA

* HANDLE SPECIAL CASES.
      ULPS = 0.0
      IF (X .EQ. Y) RETURN
* PREPARE TO RETURN A HUGE VALUE IF X AND Y HAVE
* OPPOSITE SIGN.
      ULPS = 1.0 / EPS
      IF (X .EQ. 0.0D0) RETURN
      ABSX = ABS(X)
      IF ((X / ABSX)*Y .LE. 0.0D0) RETURN

* BRACKET BSUPE1 = BETA**(E-1) .LE. ABS(X) .LT. BETA**E =
* BSUPE.
      BSUPE = 1.0
10    IF (BSUPE .LE. ABSX) THEN
          BSUPE = BSUPE*EBETA
          GO TO 10
      END IF
      BSUPE1 = BSUPE / EBETA
20    IF (BSUPE1 .GT. ABSX) THEN
          BSUPE = BSUPE1
          BSUPE1 = BSUPE / EBETA
          GO TO 20
      END IF

* GET SPACING BETWEEN MACHINE NUMBERS IN THE RANGE
* BSUPE1 TO BSUPE.
      SPACE = BSUPE1*EPS

      ABSY = ABS(Y)
      IF (ABSY .LT. BSUPE1) THEN
          ULPS = ((ABSX - BSUPE1) + (BSUPE1 - ABSY)*EBETA) / SPACE
      ELSE IF (ABSY .GT. BSUPE) THEN
          ULPS = ((BSUPE - ABSX) + (ABSY - BSUPE) / EBETA) / SPACE
      ELSE
*         NO POWER OF BETA LIES BETWEEN X AND Y SO...
          ULPS = ABS(X - Y) / SPACE
      END IF
      RETURN
      END
```

Sec. 2.3   Ulps

When run with the test driver,

```
        DOUBLE PRECISION X
        REAL SHORT, ULPS
10      READ *, X
        IF (X .EQ. 0.0D0) STOP
        SHORT = X
        PRINT *, ULPS(X, DBLE(SHORT))
        GO TO 10
        END
```

ULPS should produce a sequence of values between 0 and 1. (In fact, the values should not exceed 0.5 if the machine uses proper rounding, instead of chopping, to convert double precision numbers to single precision.) If X and Y are single precision variables with close values, then

```
        PRINT *, ULPS(DBLE(X), DBLE(Y))
```

should produce a small nonnegative integer.

PROGRAMMING ASSIGNMENT

Implement *ULPS* on your computer and check that it performs properly.

EXERCISES

1. Ignoring the fact that *ULPS* will be executed using fixed-precision arithmetic, explain intuitively how *ULPS* works. [*Hint*: First work through an example with $\beta = 10$, $p = 3$ ($p = 6$ for double precision), $x = 0.9978$, and $y = 1.0112$. Then explain the general case by expanding the following observation. Suppose that $\beta^{e-1} \le x < \beta^e \le y$ and let s denote the spacing between consecutive floating-point numbers in the range β^{e-1} to β^e, so $s = \beta^{e-1} \times \epsilon$. It follows that x and β^e differ by $(\beta^e - x)/s$ ulps. Furthermore, $\beta \times s$ is the spacing between consecutive floating-point numbers in the range β^e to β^{e+1}, so if $y \le \beta^{e+1}$, then β^e and y differ by $(y - \beta^e)/(\beta \times s)$ ulps. Why will the value computed by *ULPS* differ from $ulps(x, y)$ if two or more powers of β lie between x and y?]

2. Analyze the effects of using floating-point arithmetic to execute *ULPS*. Why are rounding errors not much of a problem? Show that if $|x| < \beta^{m+p-2}$, where m is the minimum floating-point exponent, then *ULPS* will attempt a division by zero. (Better this than a wrong result.) [*Hint*: Our assumptions about floating-point arithmetic imply that if both $\beta^{e-1} \le |x| \le \beta^e$ and $\beta^{e-1} \le |y| \le \beta^e$, then $x - y$ is computed exactly, barring underflow. Be warned that for some computers with suboptimal arithmetic this conclusion fails. Specifically, the problem arises if the machine has no "guard digit" for subtraction. See Exercise 1 of Section 3.4 for a discussion.]

2.4 UNCERTAINTY DIAGRAMS

It is sometimes appropriate to illustrate points about numerical errors not with actual numbers but rather with abstractions in which only the numbers' uncertainties and perhaps their relative magnitudes remain. Instead of saying "suppose that $x = 0.523164 \times 10^3$ approximates an unknown value lying in the range 523.1 to 523.2" we will draw a diagram like Figure 2.3, where the extent of the shaded region indicates that about one third of the digits are uncertain. In general we will have in mind a fixed floating-point number system with base β and precision p and will consider the kth of the p digits to be uncertain if the number could be in error by as much as one ulp of k-digit arithmetic. (In algebraic terms, x is an approximation to some number $x + \delta$, where $|\delta|$ might be as large as $\beta^{1-k}|x|$.) Thus if $x = 0.500000$ represents a value known only to lie between 0.499999 and 0.500001, then just the sixth digit of x will be considered uncertain. We will shade the kth of p equal segments of the box depicting x to show that x's k^{th} digit is uncertain. In this section we will give examples of both successes and failures of reasoning based on these diagrams to understand error propagation.

Figure 2.4 represents our assumption that if the operands are considered exact, then the result of a floating-point operation is in error at most half an ulp (assumption 1 of Section 2.1).

The propagation of errors when a single multiplication is applied to inexact operands can be pictured as in Figure 2.5; the picture indicates that the product is roughly as uncertain as its least certain operand. The reasoning behind this is quite simple: If you perturb one operand in, say, the third digit (from the left, ignoring leading zeros), the product will change in about its third digit (excluding carries, which do not count). Alternatively, the first two digits of the product depend on only the first two or so digits of the factors. Division follows the same rule.

For addition and subtraction the situation is more complicated. For example, when the numbers are of unequal magnitude (and more generally, when the magnitude of the sum is about that of the larger summand), we have

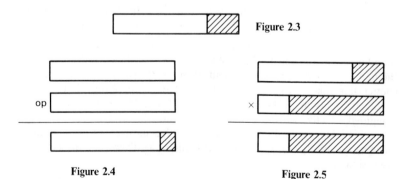

Figure 2.3

Figure 2.4

Figure 2.5

Sec. 2.4   Uncertainty Diagrams

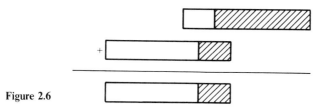

Figure 2.6

the situation shown in Figure 2.6. Unlike Figures 2.4 and 2.5, where the relative magnitudes of the numbers were irrelevant and thus were not reflected in the diagrams, in the current example the smaller summand has been shifted to make the two exponents agree. Thus the addition $0.1234 \times 10^1 + 0.3624 \times 10^3$ is diagrammed as follows:

$$0.\ 0\ 0\ \boxed{1\ 2\ 3\ 4}\ \times 10^2$$
$$+0.\boxed{3\ 6\ 2\ 4}\ \times 10^3$$

Observe that in Figure 2.6 the uncertainty of the smaller operand has a diminished influence on the sum. We see therefore that circumstances exist in which we can expect a computed value to have a greater number of accurate digits than does one of its operands.

On the other hand, the situation is reversed in the "catastrophic cancellation" pictured in Figure 2.7. There the result of a single operation on two reasonably accurate operands can be a number with few, if any, accurate digits. For instance, if only the first five figures of $x = 1.23456$ and the first four figures of $y = -1.23321$ are accurate, then the addition $x + y = 1.35000 \times 10^{-3}$ can be pictured as in Figure 2.7.

Inspection of the digit-by-digit algorithm for computing square roots makes it clear that the result of a square root operation is about as uncertain as is the operand. See Figure 2.8.

We are led to the conclusion that there is only one situation in which the result of a single operation is substantially more uncertain than are its operands, namely the occurrence of catastrophic cancellation in a $+$ or $-$ operation. For any other operation (using $+$, $-$, $\times$, $/$, or $\sqrt{\ }$) the number of uncertain digits in the result is nearly matched in one of the operands.

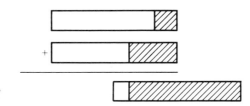

Figure 2.7

Figure 2.8

Example 1

Our first use of diagrams for analyzing error propagation is to "prove" Hypothesis 1.1a, which asserts that certain numerical errors are hard to detect with testing. The simplicity of the argument may leave the reader with the feeling that theoretical analysis provides a much sharper tool for diagnosing numerical errors than does testing. However, for reasons discussed in Chapter 3, analysis and testing should be thought of as complementing, instead of competing with, one another. In Chapter 4 we will see a case where testing can be much more efficient than analysis for uncovering errors.

Recall the following quadratic equation solver from Section 1.2:

Algorithm 1.1a

$$d \leftarrow b^2 - 4c$$

if $d \geq 0$

$$e \leftarrow \sqrt{d}$$

$$r_1 \leftarrow \frac{-b + e}{2}$$

$$r_2 \leftarrow \frac{-b - e}{2}$$

else

$$p \leftarrow \frac{-b}{2}$$

$$q \leftarrow \frac{\sqrt{-d}}{2}$$

There are only two ways in which the computed solutions [i.e., either (r_1, r_2) or (p, q)] can be inaccurate.

1. Cancellation can occur in the subtraction $b^2 - 4c$. In this case, $c \approx b^2/4$, and the roots of the quadratic nearly coincide.
2. Cancellation can occur either in the addition $-b + e$ or the subtraction $-b - e$. This happens when $|b| \approx e$, i.e., $b^2 \approx e^2 = d = b^2 - 4c$, i.e., $|c|$ is much smaller than b^2.

Sec. 2.4 Uncertainty Diagrams

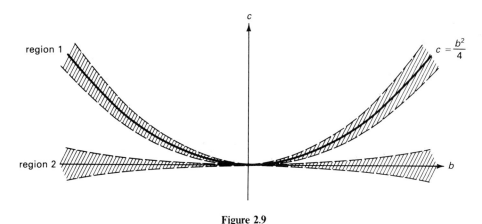

Figure 2.9

Thus the sets (b, c) of data for which Algorithm 1.1a does not satisfy Specification 1.1a are contained in the two shaded regions in Figure 2.9. Only by stumbling across a set of data in one of the two shaded regions can random testing reveal the error.

To get a more quantitative feel for the scarcity of data that reveal this error, take $\beta = 10$, $p = 6$, and $b = 0.5$. For what values of c will two or more decimal digits be lost in the subtraction $b^2 - 4c$? Such cancellation occurs only for $4c$ between 0.240001 and 0.259999, or approximately $0.06 < c < 0.065$. For instance, with $4c = 0.241234$ the subtraction $b^2 - 4c$ takes the form

$$\begin{array}{r} 0.250000 \\ -0.241234 \\ \hline 0.\underbrace{008766}_{\text{lost digits}} \end{array}$$

Thus if c were generated randomly between -1 and 1, the probability would be only $0.005/2 = 0.0025$ that such cancellation would occur.

Example 2

(Readers without a strong background in mathematics will find this example difficult. However, their perseverance will be repaid in Section 3.1.) The Taylor's series expansion of $sin(x)$, which is discussed in most calculus classes, is

$$sin(x) = x - \frac{1}{3!}x^3 + \frac{1}{5!}x^5 - \frac{1}{7!}x^7 \cdots$$

where $k! = 1 \times 2 \times \cdots \times k$. Its intuitive meaning is that given any value of x, we can (neglecting rounding errors) compute a value as close as we like to $sin(x)$ by adding up enough of the terms x, $-(1/3!)x^3, \ldots$. The following algorithm uses the fact that successive terms of this series differ by the factor $[-x^2/k(k-1)]$ for $k = 3, 5, 7, \ldots$. Terms of the series are added until a point is reached where the current term is equivalent to less than half an ulp of the sum:

$k \leftarrow 1$

$term \leftarrow x$

$sine \leftarrow x$

repeat

    $oldsine \leftarrow sine$

    $k \leftarrow k + 2$

    $term \leftarrow -term \times \dfrac{x^2}{k(k-1)}$

    $sine \leftarrow sine + term$

until $sine = oldsine$

One problem with this algorithm is that for large x the computed value of $sin(x)$ will be completely inaccurate because of rounding error. The following simple argument shows why this must be true.

First, notice that each term is computed without any floating-point additions or subtractions and hence must be accurate to within a few ulps. (A unary minus operation is presumed to be error-free.) Thus we need only worry about the propagation of errors when the terms are added.

For data $x = 20$, to take a specific example, $[x^2/k(k-1)]$ (the ratio of successive terms) exceeds 1 when $k = 3, 5, \ldots, 19$ and then falls below 1. Thus the largest term corresponds to $k = 19$ and has the value $20^{19}/19! \approx 430{,}998 \times 10^2$. It follows that in six-decimal arithmetic the procedure computes as in Figure 2.10. A one-ulp change in the largest term, say from $430{,}998 \times 10^2$ to $430{,}997 \times 10^2$, changes the sum by 100. The effect of all errors is unlikely to be *less* than this.

Example 3

These pictorial arguments suffer from the defect of sometimes producing misleading information. One way they can mislead is by making a value look more accurate than it actually is. For example, the relative error in a product of two uncertain values is roughly the sum of the relative errors in the factors, rather than the maximum of the two errors. In particular, squaring a number causes the loss of roughly one additional bit of accuracy. Thus if $y = 1.00001$ is an approximation to 1.0, then repeating the operation $y \leftarrow y^2$ gradually erodes the accuracy until, after 17 iterations, none remains. [This follows readily from the observations that $(1 + \epsilon)^2 \approx 1 + 2\epsilon$ and that $2^{17}(10^{-5}) \approx 1$.] On the other hand, our pictorial arguments wrongly suggest that the final result is still correct to within a few ulps.

The most interesting failure of this method of reasoning is that it is sometimes strikingly pessimistic. It may be the case that the errors in two operands are correlated in such a way that the *errors* cancel. A trivial example is the subtraction $x - x$: The computed result is exact regardless of the error in x. This phenomenon, that lost accuracy "mysteriously" reappears in subsequent computation, is much more subtle and important than the $x - x$ example suggests. In fact, fortunate correlation of

Sec. 2.4   Uncertainty Diagrams   41

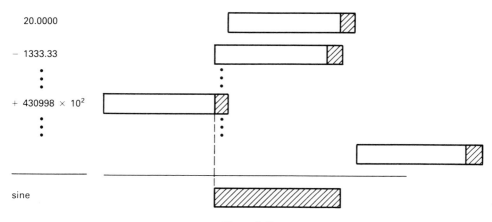

Figure 2.10

rounding errors appears in many of the best numerical methods. This is no accident: The correlation may be exactly what makes the algorithm superior.

For instance, in Chapter 3 we will see a simple case where intermediate computed values can be completely inaccurate yet lead to a final result that is correct to within a few ulps. Specifically, to evaluate $sin(t)$ we compute an integer n satisfying certain properties and then form $x = t - n\pi$. It is possible for the computed pair (n', x') to be quite different from the exact pair (n, x). However, the final result is not harmed because the errors in n and x must satisfy certain rigid conditions, namely that $n' = n \pm 1$ and $|x - x'|$ is very close to π.

Another, more complex, failure of these pictorial arguments arises in Chapter 4. If the linear system

$$0.578x + 0.323y = 0.901$$
$$0.377x + 0.212y = 0.589$$

is solved in three-digit arithmetic by the Gaussian elimination method, then the computed solution $x' = 0.441$, $y' = 2.00$ differs considerably from the exact solution $x = 1.0 = y$. However, the errors in x' and y' are correlated in such a way that

$$0.578x' + 0.323y' = 0.900898$$
$$0.377x' + 0.212y' = 0.590257$$

Simple pictorial arguments are completely inadequate to predict this fact. For instance, the *actual errors* in the addition $0.578x' + 0.323y'$ behave as shown in Figure 2.11.

James Wilkinson made the following remarks in "Some Comments from a Numerical Analyst" (*Journal of ACM*, April 1971, pp. 137–147, copyright 1971, Association for Computing Machinery, Inc.; reprinted with the permission of ACM):

> In 1946 the mood of pessimism about the stability of elimination methods for solving linear systems was at its height and was a major talking point. Bounds had been produced which purported to show that the error in the solution would be proportional to 4^n and this suggested that it would be impractical to solve

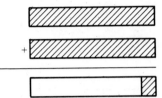

Figure 2.11

systems of even quite modest order.... I think it was true to say that at that time (1946) it was the most distinguished mathematicians who were most pessimistic, the less gifted being perhaps unable to appreciate the full severity of the difficulties. I do not intend to indicate my place on this scale, but I did find myself in a rather uncomfortable position for the following reason.

It so happens that while I was at the Armament Research Department I had an encounter with matrix computations which puzzled me a good deal.... I had been presented with a system of twelve linear equations to solve.... I finally decided to use Gaussian elimination with what would now be called "partial pivoting".... I used ten-decimal computation.... The system was mildly ill-conditioned, though we were not so free with such terms of abuse in those days, and starting from coefficients of order unity, I slowly lost figures until the final reduced equation was of the form, say,

$$.0000376235 x_{12} = .0000216312$$

At this stage I remember thinking to myself that the computed x_{12} derived from this relation could scarcely have more than six correct figures, even supposing that there had been no buildup of rounding errors.... I computed all the variables to ten figures....

Then... I substituted my solution in the original equations to see how they checked.... To my astonishment the left-hand side agreed with the given right-hand side to ten figures, i.e. to the full extent of the righthand side.... I was completely baffled by this.... I felt sure that none of the variables could have more than six correct figures and yet the agreement was as good as it would have been if I had been given the exact answer and had then rounded it to ten figures. However, the war had still to be won, and it was not time to become introspective about rounding errors....

In a series of papers and books that began appearing around 1960 Wilkinson explained why it is that the errors in the solution computed by Gaussian elimination with partial pivoting, though they may be huge, are always correlated in such a way that the given linear equations are very nearly satisfied. To verify these explanations, it was necessary to invent *backward error analysis*, a technique of analyzing the propagation of rounding errors by showing that the computed solution is the exact solution for some nearby set of data. Suffice it to say that Wilkinson's arguments are quite difficult.

EXERCISES

1. (a) Show that $\left(\sum_{i=1}^{n} x_i^2\right)^{1/2}$ is computed accurately unless underflow or overflow occurs.
   (b) Consider a floating-point number system with $\beta = 10$, $p = 3$, and $m = -100$. Find machine numbers x and y for which the computed value of $(x^2 + y^2)^{1/2}$ is incorrect by a factor of nearly $\sqrt{2}$.

2. (a) Sketch the region of the (b, c) plane where Algorithm 1.1b fails to satisfy Specification 1.1a.
   (b) Let $c = \frac{1}{8}$ and suppose that b is generated randomly between -1 and 1. What is the probability that two or more digits will cancel in the subtraction $b^2 - 4c$? [Hint: $\sqrt{0.51} \approx 0.714$.]

3. The series expansion

$$e^x = 1 + x + \frac{x^2}{2!} + \frac{x^3}{3!} + \cdots$$

suggests the following algorithm for computing the exponential function e^x:

$k \leftarrow 0$

$term \leftarrow 1$

$exp \leftarrow 1$

repeat

$\quad oldexp \leftarrow exp$

$\quad k \leftarrow k + 1$

$\quad term \leftarrow term \times \dfrac{x}{k}$

$\quad exp \leftarrow exp + term$

until $exp = oldexp$

(a) Use pictorial arguments to show that the algorithm is fairly accurate for $x \geq 0$ but generally disastrous for $x < 0$.
(b) Show that computed results will always be fairly accurate if arguments $x < 0$ are handled as follows: Set $x \leftarrow -x$, compute exp as above, and return the value $1/exp$.

4. Show that the solution of
$$0.90x + 0.34y = 0.68$$
$$0.54x + 0.21y = 0.42$$
computed by Gaussian elimination in two-digit arithmetic is $x' = 0.38$ and $y' = 1$. (Multiply the first equation by 0.6 and subtract the result from the second equation. This gives $y = 1$. Now solve the first equation for x.) Are these values near the true solution? Do they almost satisfy the equations?

3

COMPUTATION OF THE SINE FUNCTION

The main goal of this chapter is to expose certain relationships between algorithm analysis and program testing. To do this, we will consider one of the relatively few numerical algorithms whose behavior can be fully understood and get as accurate an idea as time permits of the nature of the analysis that provides this understanding. As we will see, theoretical analysis plays the following roles:

1. Analysis of error propagation in a numerical algorithm may be needed to discover an appropriate specification to be used for testing an implementation of the algorithm.
2. Analysis can suggest improvements to the algorithm that would not be discovered by blind testing.
3. Analysis of the propagation of errors in the testing procedure may be needed to make the test reliable.

On the other hand, testing is required because algorithm analysis, by itself, is not entirely adequate. One reason is that the components of an actual computing environment, including the arithmetic unit, the compiler, and the operating system, are so complex that only a few idealized characteristics of their behavior enter into an analysis. Thus conclusions, though properly deduced from the assumptions, can turn out to be invalid in practice. For example, a student trying to compute the underflow threshold σ found out that his "correct" program failed because his computer responded to underflow by returning a value near the *overflow* threshold. Also, our Experiment 4.1a had to be repeated when we noticed that the results changed because the programs were compiled so as to check for out-of-bounds array subscripts. (An unexpected side effect of our compiler's option to check subscripts is that certain intermediate values are computed in double precision.) Other examples of Murphy's law of floating-point arithmetic are mentioned in Section 2.2.

Another reason why algorithm analysis must be supplemented by program testing is that we humans are fallible, both in analyzing algorithms (even using oversimplified assumptions) and in implementing algorithms as computer programs. This point is amply illustrated by Susan Gerhart and Lawrence Yelowitz in "Observations of Fallibility in Applications of Modern Programming Methodologies" (*IEEE Transactions on Software Engineering*, Sept. 1976, pp. 195–207). If you have the time, read the entertaining and thought-provoking condemnation of formal program verification by Richard De Millo, Richard Lipton, and Alan Perlis in "Social Processes and Proofs of Theorems and Programs" (*Communications of ACM*, May 1979, pp. 271–280).

More specifically, relationships between analysis and testing are examined in the context of computing the sine function. The first two sections of this chapter present a rudimentary procedure for computing $sin(t)$ and discuss an implementation issue. The remaining sections illustrate the three roles of analysis mentioned above.

In Section 3.3 we consider the problem of testing a sine procedure using the specification that the computed value should lie within $\beta|t/sin(t)|$ ulps of the exact value $sin(t)$. An informal justification of the appropriateness of this specification is given. It is doubtful whether the specification would even come to mind without some understanding of such a justification.

In Section 3.4 we discuss the importance of attaining greater accuracy in the computed value of $sin(t)$ and present the means for doing so. The informal error analysis given in Section 3.3 indicates that all but one or two of the $\beta|t/sin(t)|$ ulps of potential error arise in the "argument reduction" $x \leftarrow t - n\pi$. It follows that the only possibility for significant gains in accuracy is to increase the precision with which π is represented and exploit this additional precision in the argument reduction. Section 3.4 shows how this improvement can be won without the use of higher-precision floating-point hardware.

Proper testing of a high-accuracy sine procedure, especially one written in the highest available floating-point precision, requires an understanding of the propagation of errors in the testing procedure. In Section 3.5 we will show how to ensure that rounding errors in the test harness will not masquerade as inaccuracy of the sine procedure.

Our choice of material, much of which is taken directly from *Software Manual for the Elementary Functions* by William J. Cody, Jr. and William Waite (Prentice-Hall, Englewood Cliffs, N.J., 1980), reflects our belief that the most important property of a sine procedure is reliability—the procedure should compute accurate values of $sin(t)$ for all legitimate arguments t and flag the others. We feel that efficiency is a less important issue.

3.1 A RUDIMENTARY SINE PROCEDURE

The sine function, $sin(t)$ for t in radians, has the graph shown in Figure 3.1. The fact that

$$\cdots = sin(t - 2\pi) = -sin(t - \pi) = sin(t) = -sin(t + \pi)$$
$$= sin(t + 2\pi) = \cdots$$

suggests that $sin(t)$ be evaluated in three steps:

1. (argument reduction step) Shift t an integer multiple of π by computing $x \leftarrow t - n\pi$, where n is chosen so that $-(\pi/2) \leq x \leq \pi/2$. Since finding the n that minimizes $t - n\pi$ is equivalent to minimizing $(t - n\pi)/\pi = t\pi^{-1} - n$, n is the integer closest to $t\pi^{-1}$. (If $t\pi^{-1}$ is midway between two integers, then either one can be chosen to be n.)
2. Compute an approximation to $sin(x)$ by applying certain elementary operations to x and to certain constants.
3. If n is even, return $sin(x)$. Otherwise return $-sin(x)$.

The second step requires elaboration. The Taylor's series expansion of $sin(x)$, which is discussed in most calculus classes, is

$$sin(x) = x - \frac{1}{3!}x^3 + \frac{1}{5!}x^5 - \frac{1}{7!}x^7 \cdots$$

where $k! = 1 \times 2 \times \cdots \times k$. In other words, the sequence of polynomials

$$p_0(x) = x$$

$$p_1(x) = x - \frac{1}{3!}x^3$$

$$p_2(x) = x - \frac{1}{3!}x^3 + \frac{1}{5!}x^5$$

etc.

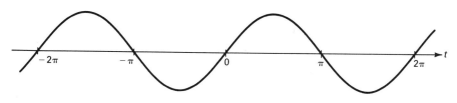

Figure 3.1

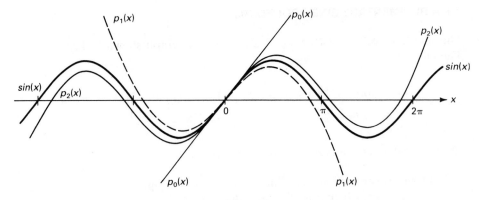

Figure 3.2

converges to $sin(x)$. See Figure 3.2. For any given precision of computer arithmetic, we can pick the smallest k such that $p_k(x)$ and $sin(x)$ differ by at most half an ulp for all x satisfying $|x| \le \pi/2$, and let the second step of the sine procedure consist of evaluating $p_k(x)$. For example, p_4 is adequate on a five-digit decimal machine (i.e., $\beta = 10$ and $\epsilon = 10^{-4}$) since if $|x| \le \pi/2$, then an additional term would satisfy

$$\left| \frac{1}{(2k+1)!} x^{2k+1} \right| = \left| \frac{1}{(2k+1)!} x^{2k} \right| \times |x|$$

$$\le \frac{1}{11!} \times \left(\frac{\pi}{2}\right)^{10} \times |x|$$

$$< \tfrac{1}{2} \times 10^{-5} \times |sin(x)|$$

$$= \tfrac{1}{2} \beta^{-1} \epsilon |sin(x)|$$

whereas the most important fact about floating-point number systems (see p. 19) shows that the spacing between adjacent floating-point numbers around $sin(t)$ is at least $\beta^{-1} \epsilon |sin(t)|$. Thus all digits of the term would be lost in the addition; see Figure 3.3.

The approximating polynomial $p_k(x)$ can be evaluated quite efficiently. The coefficients $-1/3!$, $1/5!$, $-1/7!$, ... can be computed before the

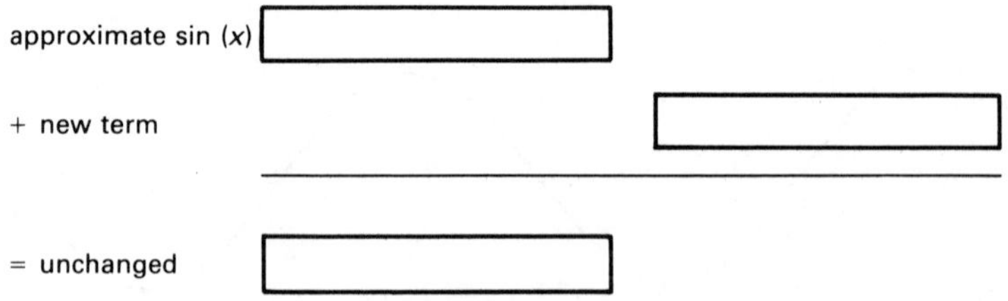

approximate sin (x) [_____]

+ new term   [_____]

= unchanged [_____]

Figure 3.3

Sec. 3.1 A Rudimentary Sine Procedure

program is written and stored as constants, so $p_k(x)$ has the form $x + c_1 x^3 + c_2 x^5 + \cdots + c_k x^{2k+1}$. To evaluate $p_k(x)$, we can compute $y \leftarrow x^2$, $r \leftarrow c_1 y + c_2 y^2 + \cdots + c_k y^k$ and $p \leftarrow x + x \times r$. Unnecessary multiplication operations can be avoided by using Horner's rule to compute r. For instance, if $k = 4$, then compute

$$r \leftarrow (((c_4 \times y + c_3) \times y + c_2) \times y + c_1) \times y$$

The polynomials $p_k(x)$ approximate $sin(x)$ much more closely around $x = 0$ than they do for x near $\pm(\pi/2)$. It is advantageous to alter the exact coefficients c_i slightly so as to distribute the error more evenly, since by doing so it is possible to achieve the same accuracy with a lower-degree polynomial. Cody and Waite give the following coefficients, where b is the number of bits in the host machine's floating-point fraction. [The method of computing these coefficients is beyond the scope of this book. The interested reader can consult *Computer Evaluation of Mathematical Functions* by C. T. Fike (Prentice-Hall, Englewood Cliffs, N.J., 1968).]

For $b \leq 24$, set $k = 4$ and

$$c_1 = -0.16666\ 65668$$
$$c_2 = 0.83330\ 25139 \times 10^{-2}$$
$$c_3 = -0.19807\ 41872 \times 10^{-3}$$
$$c_4 = 0.26019\ 03036 \times 10^{-5}$$

For $25 \leq b \leq 32$, set $k = 5$ and

$$c_1 = -0.16666\ 66660\ 883$$
$$c_2 = 0.83333\ 30720\ 556 \times 10^{-2}$$
$$c_3 = -0.19840\ 83282\ 313 \times 10^{-3}$$
$$c_4 = 0.27523\ 97106\ 775 \times 10^{-5}$$
$$c_5 = -0.23868\ 34640\ 601 \times 10^{-7}$$

For $33 \leq b \leq 50$, set $k = 7$ and

$$c_1 = -0.16666\ 66666\ 66659\ 653$$
$$c_2 = 0.83333\ 33333\ 27592\ 139 \times 10^{-2}$$
$$c_3 = -0.19841\ 26982\ 32225\ 068 \times 10^{-3}$$
$$c_4 = 0.27557\ 31642\ 12926\ 457 \times 10^{-5}$$
$$c_5 = -0.25051\ 87088\ 34705\ 760 \times 10^{-7}$$
$$c_6 = 0.16047\ 84463\ 23816\ 900 \times 10^{-9}$$
$$c_7 = -0.73706\ 62775\ 07114\ 174 \times 10^{-12}$$

For $51 \leq b \leq 60$, set $k = 8$ and

$c_1 = -0.16666\ 66666\ 66666\ 65052$

$c_2 = 0.83333\ 33333\ 33316\ 50314 \times 10^{-2}$

$c_3 = -0.19841\ 26984\ 12018\ 40457 \times 10^{-3}$

$c_4 = 0.27557\ 31921\ 01527\ 56119 \times 10^{-5}$

$c_5 = -0.25052\ 10679\ 82745\ 84544 \times 10^{-7}$

$c_6 = 0.16058\ 93649\ 03715\ 89114 \times 10^{-9}$

$c_7 = -0.76429\ 17806\ 89104\ 67734 \times 10^{-12}$

$c_8 = 0.27204\ 79095\ 78888\ 46175 \times 10^{-14}$

In summary, our rudimentary sine procedure is as follows:

Algorithm 3.1

..Step 1. Argument reduction:

$n \leftarrow$ the integer closest to t/π

$x \leftarrow t - n\pi$

..Step 2. Evaluate the approximating polynomial:

$y \leftarrow x^2$

$r \leftarrow (\cdots(c_k \times y + c_{k-1}) \times y \cdots) \times y$

$p \leftarrow x + x \times r$

..Step 3. Restore the sign:

if n is even

$sine \leftarrow p$

else

$sine \leftarrow -p$

Typographical Mistakes. It seems plausible that any blunder committed when implementing Algorithm 3.1 stands a good chance of being caught if the program is run on an angle t satisfying $|t| > \pi/2$ (to check the argument reduction step), an angle t satisfying $|t| < \pi/2$ (to check the polynomial approximation), and an angle t for which n is odd (to check the sign restoration). As it turns out, there are additional considerations that determine which collections of angles are good for catching programming mistakes, but in any case we can consider the following experimental hypothesis.

Sec. 3.1 A Rudimentary Sine Procedure

Hypothesis 3.1
To expose almost any typographical mistake in an implementation of Algorithm 3.1 it is sufficient to execute the procedure on a few carefully chosen angles t.

We performed a mutation experiment (as defined in Section 1.1) using (1) an implementation of Algorithm 3.1, (2) the angles 1.5, 10, and -10, and (3) the acceptance criterion that the computed result should differ by less than 10^{-6} from the value computed by FORTRAN's *SIN* function. (On our machine, $10^{-6} \approx 8\epsilon$.) All typographical mistakes were exposed; i.e., no mutant survived. On the other hand, each of the following sets of angles proved to be inadequate to catch all typographical mistakes.

1.5 and 10
1.5 and -10
1.5, 5, and -5
1.0, 10, and 10

(In Exercise 1 of Section 3.3 you will see why 1.5 is better than 1.0 for debugging implementations of Algorithm 3.1.)

It is worth noting that the "typographical mistakes" entering into the experiment did not include minor mistakes in the constants c_i. (See the definition of *typographical change* on p. 3.) Check your program's constants carefully!

PROGRAMMING ASSIGNMENTS

1. Write a sine procedure and check that it works for a few angles t. The following constants may prove useful:

$$\pi = 3.14159\ 26535\ 89793\ 23846\ldots$$

$$\pi^{-1} = 0.31830\ 98861\ 83790\ 67154\ldots$$

(The rules of FORTRAN allow you to include these blanks in your constants, and we suggest that you do so.)

2. (Optional) Determine the differences, in terms of execution efficiency for a particular computer and compiler, among
   (a) Keeping the c_i as simple variables $C1, C2, \ldots$ and evaluating

   $$r \leftarrow (((C4 \times y + C3) \times y + C2) \times y + C1) \times y$$

   (b) Keeping the c_i in an array $C[i]$ and evaluating

   $$r \leftarrow (((C[4] \times y + C[3]) \times y + C[2]) \times y + C[1]) \times y$$

(c) Keeping the c_i in an array $C[i]$ and evaluating

$r \leftarrow 0.0$

for $i = 4$ down to 1

$r \leftarrow (r + C[i]) \times y$

If possible, determine the answer by inspecting listings of the compiled program instead of by timing execution. (This assignment is included because we believe that the student should get some practice at deciphering assembler language listings produced by compilers.)

EXERCISES

1. Do the values computed by your sine procedure satisfy $sin(-t) = -sin(t)$ for all machine numbers t? (This property is desirable because users will expect it to hold and may use the sine procedure in a way that fails if the property does not hold. We recommend writing your sine procedure so that a negative argument t is immediately replaced by $-t$, with final adjustment of the sign of the value to be returned. Doing so makes it easy to answer this exercise affirmatively.)
2. (Optional) Show that no accurate sine procedure can satisfy the following requirement: *If x and y are machine numbers such that $-(\pi/2) \leq x < y \leq \pi/2$, then the computed values satisfy $sin(x) < sin(y)$.* [*Hint*: The sine function maps the interval $[-(\pi/2), \pi/2]$ into the smaller interval $[-1, 1]$.]

3.2 IMPLEMENTATION ISSUES: SMALL AND LARGE ANGLES

In this section we will give reasons why a sine procedure should treat very large initial angles t and very small reduced angles x as special cases. In particular, we recommend that the argument reduction step of Algorithm 3.1 be augmented to take the following form:

if $|t| \geq \Theta$

print a message that $sin(t)$ is poorly determined

and/or terminate execution

$n \leftarrow$ the integer closest to t/π

$x \leftarrow t - n\pi$

if $|x| \leq \theta$

sine $\leftarrow x$

return

Criteria for determining the thresholds Θ and θ will be discussed.

The Need for Special Handling of Small and Large Angles.

There are two distinct constraints on the sizes of the initial angles t for which a sine procedure can compute meaningful values. (A third constraint will be discussed in Section 3.4.)

1. As t increases in magnitude, so does the spacing between t and the next larger machine number. At some point a request to compute $sin(t)$ should be eyed with suspicion, since even the slightest uncertainty in t will mean that $sin(t)$ is poorly determined. For instance, if the distance from t to the next machine number exceeds π, then even a one-ulp uncertainty in t leaves $sin(t)$ completely uncertain. We will see in the next section that, more generally, a one-ulp uncertainty in t leaves $sin(t)$ uncertain by about $|t/sin(t)|$ ulps.

2. In cases where a floating-point number is allotted more bits of storage than is a machine integer, an attempt to compute n may provoke an *integer* overflow.

The sine procedure should flag any t for which it cannot produce a useful value.

Small reduced angles x are less of a problem. For instance, underflow poses no serious difficulty on those machines that silently replace an underflowed value by zero [so long as the procedure computes $sin(x)$ as x plus a correction term]. However, even for such machines it is best to explicitly approximate $sin(x)$ with x whenever x is so small as to be the closest floating-point number to the value $p_k(x)$ of the approximating polynomial. Programs written this way avoid expending unnecessary execution time computing $p_k(x)$ for tiny x, avoid costly operating system responses to underflow, and enjoy the benefit of *obviously* working as expected on small arguments. No time need be spent wondering whether, because of rounding errors, there exists a tiny x such that the computed $sin(x)$ differs from x.

If ϵ denotes machine epsilon, then the sine procedure can immediately return x as its value if $|x| \leq \sqrt{\epsilon}$. To see why this is so, notice that $y = x^2$ then satisfies $0 \leq y \leq \epsilon$, that $r = c_1 y + c_2 y^2 + \cdots + c_k y^k$ (very nearly) satisfies $-(\epsilon/6) \leq r \leq 0$ since $c_1 \approx -\frac{1}{6}$, and that $p = x + x \times r$ satisfies $|p - x| = |xr| \leq (\epsilon|x|)/6$. The most important fact about floating-point number systems implies that the spacing between representable numbers near x is at least $(\epsilon|x|)/\beta$, so x approximates p to within the machine's precision.

For somewhat more concrete examples of how one might set a sine procedure's cutoff points for "very large" angles t and "very small" angles x, let us imagine a computer with 24-bit single precision fractions, 54-bit double precision fractions, and 32-bit integers. Because one bit of an integer gives the sign, the largest representable integer is $2^{31} - 1$.

Example 1: A Single Precision Sine Procedure

Constraint 1. The 24-bit precision is roughly equivalent to seven decimal digits. As mentioned above, a one-ulp uncertainty in t results in an uncertainty in $sin(t)$ of about $|t/sin(t)| \geq |t|$ ulps, so if $t \geq 10{,}000$, then at most the first two digits of $sin(t)$ are determined. For example, changing t from 12345.67 to 12345.68 changes $sin(t)$ from $-0.7097\ldots$ to $-0.7026\ldots$. Even if t is known and is exactly representable as a floating-point number, the error in $n\pi$ (caused by the error in representing π and the rounding error in the floating-point multiplication) will be of this same amount and will affect the computed value of $sin(t)$ similarly. Thus the program might well begin with a statement "if $|t| > 10^4$ then write $\ldots$," producing a message that $sin(t)$ is poorly determined. A conservative approach would be to also terminate execution under that condition. However, any choice of a cutoff point will be essentially arbitrary and will displease some users.

Constraint 2. This only limits t to roughly $|t| < \pi \times 2^{31} \approx 6.7 \times 10^9$, so it is far less restrictive than is constraint 1. However, in this case it is quite clear that the sine procedure should terminate execution when faced with an angle t above this threshold.

Very Small Angles. Single precision machine epsilon on our hypothetical computer is $\epsilon = 2^{-23}$, so $\sqrt{\epsilon}$ is about 3.5×10^{-4}. Thus the single precision sine procedure might as well just immediately return the value x if $|x| < 3.5 \times 10^{-4}$.

Example 2: A Double Precision Sine Procedure

Constraint 1. This constraint causes little difficulty since an uncertainty of an ulp in t is unlikely to influence the first five decimal digits of $sin(t)$ unless $|t| \geq 10^{10}$. The reasoning is as follows. Knowing t to 54 bits is roughly equivalent to knowing 16 decimal digits. If $|t| < 10^{10}$, then at most 11 of these digits can cancel in forming x, unless $|x| < 0.1$. The remaining five or more digits of x determine the value of the sine function to a like number of digits.

Constraint 2. As for single precision, this requires $|t| < \pi \times 2^{31}$. Thus the initial check might be "if $|t| > 6.7 \times 10^9$ then $\ldots$."

Very Small Angles. For double precision on our hypothetical computer, $\epsilon = 2^{-53}$, so the cutoff point for small x should be about $2^{-26.5}$, i.e., about 10^{-8}.

PROGRAMMING ASSIGNMENT

Modify your sine procedure so that it handles small angles and large angles appropriately.

3.3 TESTING YOUR SINE PROCEDURE

The purpose of this section is to justify the use of the following specification for sine procedures:

Specification 3.1

The computed value of $sin(t)$ should lie within $\beta |t / sin(t)|$ ulps of the true value, except perhaps in cases where $|sin(t) / t|$ is not much bigger than machine epsilon.

Specification 3.1 leads immediately to the following procedure for testing sine functions.

Test 3.1

$\beta \leftarrow$ the floating-point base

$\epsilon \leftarrow$ machine epsilon

repeat until satisfied

    $t \leftarrow$ a random angle

    true $\leftarrow$ the true $sin(t)$ determined by some means

    if $|true / t| \leq 10\epsilon$ ..the "10" is more or less arbitrary

        report that excessive cancellation invalidates this t

    else

        sine $\leftarrow sin(t)$ ..as computed by the method being tested

        $\omega \leftarrow$ ulps(*sine*, *true*) $\times |true / t|$

        if $\omega > \beta$

            report that Specification 3.1 is violated

            terminate the test

    report that Specification 3.1 appears to be met

The analysis of Algorithm 3.1 is difficult, in spite of our attempts to simplify it as much as possible. But how else would you discover Test 3.1?

Discussion of Specification 3.1. For arguments t around 0, $|t/sin(t)| \approx 1$. As t varies from $n\pi$ to $(n + 1)\pi$, $|sin(t)|$ varies from 0 up to 1 and back down to 0, so if n is neither -1 nor 0, then $|t/sin(t)|$ varies from ∞ down to $|t|$ and back up to ∞. Thus, ignoring the factor β, Specification 3.1 requires that the maximum error in the computed value of $sin(t)$ behave roughly like the graph in Figure 3.4. The ∞-ulp error occurs at the *computed* value of $n\pi$. When this value is taken as t, the computed reduced argument $t - n\pi$ and the computed $sin(t)$ are zero, whereas the true $sin(t)$ is nonzero. (The computed $n\pi$, being a floating-point number and hence a rational number, cannot equal the exact $n\pi$.)

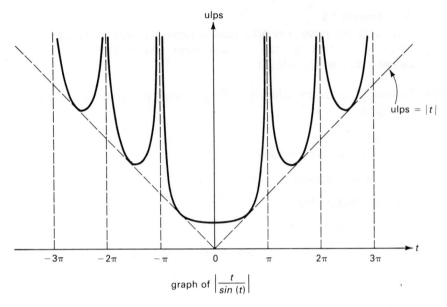

Figure 3.4

Our informal error analysis will turn up some useful facts that can be summarized with the help of the following notation:

p the approximating polynomial
x the true reduced argument
X the computed reduced argument
$p(X)$ the exact value of p at X
$P(X)$ the computed value of p at X; with restoration of the sign, this is the computed value of $sin(t)$

The difference between the true value of $sin(t)$ and the computed value is $sin(x) - P(X)$. Writing

$$sin(x) - P(X) = (sin(x) - p(x)) + (p(x) - p(X)) + (p(X) - P(X))$$

breaks the error into three components with the following properties:

1. $sin(x) - p(x)$. This is the difference between the sine function and the approximating polynomial. It is negligible compared with the other components of the error.
2. $p(x) - p(X)$. This is the difference between the values of the approximating polynomial at x and at X, which arises because of errors in computing $t - n\pi$. It should be less than $\beta |t/sin(t)|$ ulps.

Sec. 3.3 Testing Your Sine Procedure

3. $p(X) - P(X)$. This is the error caused by use of inexact arithmetic to evaluate the approximating polynomial at X. It should not exceed an ulp or two.

Properties 2 and 3 will be verified in this section. Our analysis will treat the errors in the order they are committed by the sine procedure. We will consider the effects of errors in the machine representations of π, π^{-1}, and the c_i, as well as rounding errors in the arithmetic operations.

Computation of n. The value $t\pi^{-1}$ is computed to within an ulp or so, and it is possible for this tiny error to result in an incorrect value of n. For example, if $t = 7.85$, then $t\pi^{-1}$ is computed in three-digit arithmetic as $7.85 \times 0.318 = 2.50$, whereas the exact value is $2.498\ldots$. This could result in a computed $n = 3$, whence $t - n\pi = -1.574\ldots < -1.570\ldots = -(\pi/2)$, while the correct value $n = 2$ leads to a reduced argument $x \approx \pi/2$. In general the only effect of an incorrect value of n is that the true x might be within an ulp or so of one of the numbers $-(\pi/2)$ or $\pi/2$, while the reduced argument found with the computed n lies near the other. This is of no consequence since the approximating polynomial will still be valid at the incorrect reduced argument. For the rest of the chapter we will use n to denote the value that is actually computed.

Argument Reduction. Throughout the discussion we will assume that the computed values X and $P(X)$ are not completely inaccurate, so approximations like $X \approx x$ are valid for order-of-magnitude estimations. [Indeed, the reason that Specification 3.1 excuses the case $|sin(t)/t| \approx \epsilon$ is to exclude consideration of situations in which all, or nearly all, digits cancel in the subtraction $t - n\pi$.]

The subtraction $t - n\pi$ appears as shown in Figure 3.5. The uncertainty in $n\pi$, which is caused by rounding π to a machine number and by the floating-point multiplication, should be roughly an ulp.

Let the computed value of $n\pi$ be $n\pi - e$, so that e is roughly the spacing between adjacent floating-point numbers in the vicinity of $n\pi$ and t. Because of cancellation, the subtraction is exact for the given operands, so $X = t - (n\pi - e) = (t - n\pi) + e = x + e$. In going from t to x, we have

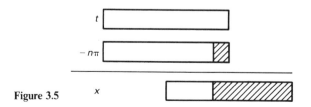

Figure 3.5

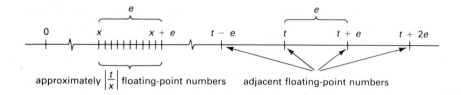

Figure 3.6

decreased the magnitudes by a factor $|x/t|$ and hence decreased the spacing between nearby floating-point numbers by about the same factor. (Notice that dividing a machine number by the floating-point base β decreases the spacing to the next machine number by *exactly* the factor β^{-1}.) Thus $X = x + e$ is about $|t/x|$ ulps away from x.

Figure 3.6 may clarify the situation. Let t and $t + e$ be adjacent floating-point numbers and scale t by the factor x/t to get x. Then there are about $|t/x|$ floating-point numbers between x and $x + e$.

For a safe upper limit to the ulps difference between x and $x + e$ we need to introduce a factor β, which yields the bound of $\beta|t/x|$ ulps. The reason for this extra factor is to account for the fact that a tiny change in x, while making almost no difference in $|t/x|$, can decrease the spacing between adjacent floating-point neighbors of x by the factor β. In particular, this happens when $|x|$ lies just above some power of β.

A more rigorous verification that this factor β does the job follows from the most important fact about floating-point number systems. The spacing s between representable numbers around x satisfies $s \geq \beta^{-1}\epsilon|x| = \beta^{-1}|x/t| \times (\epsilon|t|)$, so $\epsilon|t| \leq \beta|t/x|s$. If e is the spacing between t and an adjacent floating-point number, then $|e| \leq \epsilon|t| \leq \beta|t/x|s$. Thus $|x - (x + e)| = |e| \leq \beta|t/x|s$, so x and $x + e$ differ by at most $\beta|t/x|$ ulps.

Evaluation of p. Let $p(x) = x + c_1x^3 + c_2x^5 + \cdots + c_kx^{2k+1}$, and let the uncertainty in X be depicted as shown in Figure 3.7. According to the discussion in Section 2.4, the uncertainty diagrams for the terms $c_1X^3, \ldots, c_kX^{2k+1}$ are shaded to about the same extent. Since $x^2 < 2.5$ and the coefficients c_i decrease quickly in size (as i increases), it is not hard to see that the terms decrease quickly in size.

We offer the following formulas for the readers who find such things comforting. (Others can skip to the next paragraph.) Since $|c_i| \approx 1/(2i + 1)!$,

Figure 3.7

Sec. 3.3 Testing Your Sine Procedure

the ratios of successive terms of p satisfy

$$\left|\frac{c_i x^{2i+1}}{c_{i-1} x^{2i-1}}\right| = \left|\frac{c_i x^2}{c_{i-1}}\right|$$

$$\approx \frac{(2i-1)!}{(2i+1)!} x^2$$

$$= \frac{1}{2i(2i+1)} x^2$$

$$< \frac{2.5}{2i(2i+1)}$$

for $i = 2, 3, \ldots, k$. For $i = 1$ we have

$$|c_1 x^3| < \frac{2.5}{6} \times |x| < \frac{1}{2}|x|$$

and later terms decrease at an even faster rate.

The rapid decrease in the sizes of the terms means that their sum $p(X)$ appears as shown in Figure 3.8. Since X was shown to fall within $\beta|t/x|$ ulps of x, $p(X)$ should agree with $p(x)$ to within $\beta|t/x|$ ulps, which is approximately (though less than) $\beta|t/sin(x)| = \beta|t/sin(t)|$ ulps. This confirms property 2, mentioned above.

Virtually the same argument shows that $P(X)$ is within an ulp or two of $p(X)$ (property 3, above) if we assume that p is evaluated by the naive method of computing the necessary powers of X, multiplying each by the appropriate c_i, and adding. Analysis of Horner's rule is a little harder and leads to the same conclusion. See Exercise 2.

Specifically, let us consider X to be exact. The uncertainty in a computed term $c_i X^{2i+1}$ is only a few ulps, and Figure 3.8, with the shaded regions moved way over to the right (and removed entirely from X), shows that evaluation of p is quite immune to rounding errors.

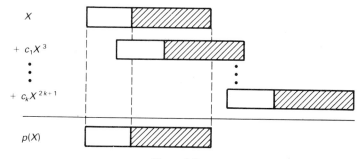

Figure 3.8

PROGRAMMING ASSIGNMENT

Apply Test 3.1 to your sine procedure. For example, you might use a random number generator to produce 1000 angles t, arranged to include small, large, positive, and negative numbers. If for some reason (e.g., that your machine chops, rather than rounds, the results of floating-point operations) $\omega(t)$ slightly exceeds β for some angles t, then formulate a revised specification that is appropriate for your machine. If some value of $\omega(t)$ is much larger than β, then locate the mistake in your program and fix it. If all values of $\omega(t)$ are less than 0.1, then there is probably something wrong with your testing procedure since even a half-ulp error in the computed value of $sin(t)$ when $|t| < \pi/2$ will score $\omega(t) = 0.5|sin(t)/t| \approx 0.5$.

EXERCISES

1. Imagine the following situation. You are testing your sine procedure with the hope of verifying that there are no mistakes in the constants c_i. You have a way of getting values of $sin(t)$ that are guaranteed to be accurate, but you can only afford to get one such value. Thus you can only test your program on one argument t. Why would a value $t \approx 1.5$ be a much better choice than $t \approx 0.1$? Also, show that the smaller coefficients c_i can have many incorrect digits without affecting the computed value of $sin(t)$ for *any* t.

2. For any function f, define the *relative derivative* of f to be
$$f^\rho(t) = f'(t)\frac{t}{f(t)}$$
where $f'(t)$ is the (ordinary) derivative of t.
   (a) Verify the approximate equality
$$\frac{f(t') - f(t)}{f(t)} \approx f^\rho(t)\frac{t' - t}{t}$$
   (b) Show that if t' and t differ by u ulps, then $f(t')$ and $f(t)$ differ by approximately $|u \times f^\rho(t)|$ ulps. Thus the absolute value of the relative derivative of f at x provides a "condition number" indicating the sensitivity of $f(x)$ to changes in x.
   (c) Show that if t' and t differ by u ulps, then $sin(t')$ and $sin(t)$ differ by at most approximately $|t/sin(t)|$ ulps.

3. (Optional and difficult) Assume that x is a number that inherits error from some previous computation, that $x^2 < 2.5$, and that c_1, c_2, c_3, and c_4 satisfy $|c_i| \approx 1/(2i + 1)!$ (or, more generally, that the c_i decrease rapidly in size as i increases). Let p be computed by the following operations:

$$y \leftarrow x^2$$
$$r \leftarrow (((c_4 \times y + c_3) \times y + c_2) \times y + c_1) \times y$$
$$p \leftarrow x + x \times r$$

Argue that x and p have roughly the same number of accurate digits.

3.4 A MORE ACCURATE SINE PROCEDURE

Increased accuracy, for large angles t, can be attained by performing a more accurate argument reduction $x \leftarrow t - n\pi$. For sine procedures working in single precision, one can compute x using double precision. This section explains why double precision sine procedures obtain the same effect as a "$2\frac{1}{2}$ precision" argument reduction if the instruction $x \leftarrow t - n\pi$ is changed to an instruction like

$$x \leftarrow (t - n \times 3.140625) - n \times 9.67653589793 \times 10^{-4}$$

The technique hinges on the fact that the product $n\pi$ involves numbers with special properties. Namely, π can be determined as accurately as desired, while n, being an integer, can usually be represented with relatively few digits.

Is Greater Accuracy Worthwhile? It can be argued that the sine procedure developed and explained above is as accurate as one might reasonably want. The discussion of property 3 in the preceding section shows that the computed value of $sin(t)$ is within about an ulp of being correct *if the computed reduced argument X is taken to be exact.* Moreover, we saw that X is equal to $t - (n\pi + e) = (t - e) - n\pi$, where e is about an ulp of $n\pi$. Since $t \approx n\pi$, it follows that the computed value is within an ulp or so of the true value $sin(T)$ for some $T = t - e$ within about an ulp of t. Thus if t is not a known machine number, then the inaccuracy due to computational errors is no worse than that due to uncertainty in the data.

On the other hand, there are reasons for wanting a more accurate sine procedure. In the first place, there exist cases where t is a known machine number, for instance, the problem of computing $sin(22)$. More important, users occasionally write programs whose correct functioning depends on certain trigonometric facts, such as $sin(2t) = 2sin(t)cos(t)$ or $|sin(t)| \leq 1$. They will be dismayed if their program fails because the classical laws are violated in some crucial way on the computer, e.g., when an attempt to take the square root of $1 - sin^2(t)$ causes their program to terminate. The easiest way to decrease the frequency and magnitude of these violations is to produce accurate procedures. (Exercise 2 of Section 3.1 shows that not all ideal properties of a sine function are attainable.)

Attaining Greater Accuracy. A single precision sine procedure running on a machine with double precision hardware can be modified to evaluate the expression $t - n\pi$ in double precision. The double precision computed value X is exactly $T - n\pi$, where T is within about one double precision ulp of the original argument t. Rounding X and evaluating the approximating polynomial p at the rounded value produces a result within about a single precision ulp of the exact $p(X)$. (These assertions follow from the discussion in the previous section.) It follows that the computed $sin(t)$ is within about a single precision ulp of $sin(T)$, where T is within about a double precision ulp of t. For example, suppose that the computer has 24-bit single precision fractions and 54-bit double precision fractions. Then the computed $sin(t)$ agrees to nearly 24 significant bits with $sin(T)$, where T agrees with t to about 54 significant bits.

What can be done if a higher precision is not conveniently available? The following discussion hinges on the assumption that if cancellation occurs when numbers t and s are subtracted, then $t - s$ is computed exactly. This condition is implied by our working hypothesis that the computed difference is the floating-point number closest to the true result, since cancellation guarantees that $t - s$ can be represented with fewer digits than are needed for t and s. However, under certain rare conditions that arise in practice the method proposed will fail (Exercise 1a), and a slightly different approach is required (Exercise 1b).

The crucial observations will be illustrated with six-digit decimal arithmetic. Once understood, they are readily generalized to other floating-point number systems. If $\pi_1 = 3.14$ and $\pi_2 = 1.59265 \times 10^{-3}$, then $\pi_1 + \pi_2$ represents 50% more of π's digits than does working precision. For $n \leq 3185$, $n\pi_1$ takes only six figures to represent exactly (because if $n < 3185$, then $n\pi_1 < 10{,}000$ and because $3185 \times \pi_1$ happens to end in a zero). Thus if $n \leq 3185$, then $n\pi_1$ and $t - n\pi_1$ are computed exactly, so evaluating $x \leftarrow (t - n\pi_1) - n\pi_2$ has the same effect as subtracting from t a nine-digit representation of $n\pi$. See Figure 3.9. For instance, consider the computation of $sin(22)$. (Since 22 is close to 7π, this provides a good example of cancellation in the subtraction $t - n\pi$.) With the above approach we would compute $(t - 21.98) - 0.0111486 = 0.88514 \times 10^{-2}$. However, for most values of $n > 3185$ the trick fails, since $n\pi_1$ is not computed exactly, and x is no more accurate than before.

The general idea is that $t - n\pi_1$ can be computed exactly if (1) the number of digits needed to represent n plus the number used for π_1 does not exceed the number p of digits in a floating-point fraction (which guarantees that $n\pi_1$ is machine-representable) and (2) cancellation occurs in the subtraction $t - n\pi_1$, so the true difference is a machine number and hence computed exactly on a machine satisfying our working assumption. Cody and Waite make the following suggestions for a b-bit floating-point fraction. For $b < 32$ the constants

$$\pi_1 = 3.11 \quad \text{(octal)}$$
$$= 3.140625 \quad \text{(decimal)}$$
$$= \frac{201}{64}$$
$$\pi_2 = 9.6765\,35897\,93 \times 10^{-4}$$

Sec. 3.4   A More Accurate Sine Procedure   63

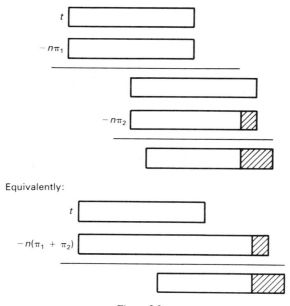

Figure 3.9

provide an extra 8 bits of precision in the argument reduction, thereby decreasing the sine procedure's error by a factor of roughly $\frac{1}{256}$ for large angles t. For $b \geq 32$ the constants

$$\pi_1 = 3.1104 \quad \text{(octal)}$$
$$= 3.14160\,15625 \quad \text{(decimal)}$$
$$= \frac{3217}{1024}$$
$$\pi_2 = -8.9089\,10206\,76153\,73566\,17 \times 10^{-6}$$

provide an extra 12 bits of precision and a 4096-fold increase in accuracy for large t. Alternate ways of obtaining the desired bit pattern (i.e., the first 8 or 12 bits of π) are given so that if, say, the language translator handles the constant *3.140625* incorrectly, one can use *201/64* (whose correct representation depends more on the accuracy of floating-point division than on the translator).

For instance, on a machine with a 54-bit floating-point fraction, this latter scheme lets us write a sine procedure that computes a value agreeing to about 54 significant bits with the exact sine of some number sharing about 66 significant bits with the given argument. The restriction that $n\pi_1$ can be computed exactly requires $|n|$ to be representable in $54 - 12 = 42$ bits, i.e., $|n| < 2^{42}$. This condition on the size of angles t that can be handled accurately is typically less restrictive than those mentioned in Section 3.2.

PROGRAMMING ASSIGNMENTS

1. Implement the above argument reduction technique.
2. (Optional) Using multiple entry points or an equivalent mechanism, write a high-accuracy procedure that evaluates both the sine and cosine functions. [*Hint*: The identity $cos(t) = sin[t + (\pi/2)]$ can be used if care is taken. For instance, the shifts $t \leftarrow t + (\pi/2)$ and $x \leftarrow (t - n\pi_1) - n\pi_2$ can be combined as follows:

   $n \leftarrow$ an integer such that $-\pi \leq t - n\pi \leq 0$

   $n_1 \leftarrow n - \frac{1}{2}$

   $x \leftarrow (t - n_1\pi_1) - n_1\pi_2$]

3. (Optional) Compute constants π_1 and π_2 that gain more than 12 bits of accuracy in the argument reduction $x \leftarrow (t - n\pi_1) - n\pi_2$. [*Hint*: The octal representation of π begins

   3.11037 55242 10264 30215 14230 63050 56006 70163]

EXERCISES

1. The first step in the subtraction of floating-point numbers is to shift the fraction of the smaller number far enough to the right to compensate for any difference in exponents. For instance, the subtraction $1.01 - 0.983$ would begin with the following shift:

$$\begin{matrix} 10^1 \times 0.101 \\ 10^0 \times 0.983 \end{matrix} \Rightarrow \begin{matrix} 10^1 \times 0.101 \\ 10^1 \times 0.0983 \end{matrix}$$

Notice that to obtain maximum accuracy, the shifted fraction needs to be held in a register with an extra digit of precision. Without this *guard digit for subtraction*, the digit 3 would be lost, and the computed value would be $10^{-1} \times 0.300$ in spite of the fact that the exact result $10^{-1} \times 0.270$ is machine-representable.

Suppose the argument reduction step $x \leftarrow (t - n\pi_1) - n\pi_2$ is used on a computer with no guard digit for subtraction.

(a) Show that the reduction works properly as long as one of the following conditions holds:

   (i) No integer power β^k of the floating-point base lies between $|t|$ and $|n\pi_1|$,
   (ii) $|n\pi_1| < \beta^k \leq |t|$ for some k, but n is not huge, or
   (iii) $|t| < \beta^k \leq |n\pi|$ for some k, but the last digit in t's fraction is zero.

[*Hint*: If subtraction of nearly equal numbers involves shifting a fraction, then no error will occur if the last digit in the fraction portion of the smaller number is zero.]

(b) Let $AINT(t)$ be the floating-point representation of the integer part of t. [Thus $AINT(-3.7) = -3.0$.] Show that the argument reduction step

$$t_1 \leftarrow AINT(t)$$
$$t_2 \leftarrow t - t_1$$
$$x \leftarrow (t_1 - n\pi_1) + t_2 - n\pi_2$$

gains extra precision even on machines without a guard digit for subtraction. [*Hint*: Unless t is huge, the fraction portion of t_1 ends with a zero.]

2. (Optional) Show that extra precision is gained by the argument reduction scheme proposed in Programming Assignment 2 for the cosine procedure. [*Hint*: n_1 can be represented with only one more digit than required by n.]

3.5 TESTING HIGH-ACCURACY SINE PROCEDURES

This section presents a strategy for testing sine procedures that are designed to lose fewer digits of precision, for large angles t, than are lost if t is perturbed by one ulp. The strategy is particularly useful for procedures that operate in the highest hardware precision and use the argument reduction $x \leftarrow (t - n\pi_1) - n\pi_2$. In particular, we will motivate and justify the following test procedure:

Test 3.2

    $g \leftarrow$ the number of bits presumably gained in argument reduction

    $\beta \leftarrow$ the floating-point base

    $\epsilon \leftarrow$ machine epsilon

    repeat until satisfied

        .. Generate the basic angle.

            $m \leftarrow$ a random integer

            $z \leftarrow$ a random number satisfying $-\dfrac{\pi}{2} \leq z \leq \dfrac{\pi}{2}$

            $t \leftarrow 3m\pi + z$

        .. Purify t; set the last two bits to zero.

            $y \leftarrow \dfrac{t}{3}$

            $u \leftarrow 5y - 4y$

            $t \leftarrow 3 \times u$

        .. Apply the triple angle identity.

            $sine \leftarrow \sin(t)$

$s \leftarrow \sin(t/3)$

if $\left|\dfrac{sine}{t}\right| \leq 10\epsilon$ or $\left|\dfrac{t/3}{s}\right| \leq 10\epsilon$

   report that excessive cancellation invalidates this t

else

   $ident \leftarrow s \times (3 - 4 \times s^2)$

   $w \leftarrow ulps(sine, ident)$

   $\omega \leftarrow w / \left(2^{-g}\left|\dfrac{t}{sine}\right| + 2^{-g}\left|\dfrac{t/3}{s}\right| + 2\right).$

   if $\omega > \beta$

      report that Specification 3.2 is violated

      terminate the test

report that Specification 3.2 appears to be met

Potential Pitfalls. Our confidence in the effectiveness of reductions $x \leftarrow (t - n\pi_1) - n\pi_2$ rests on a number of idealizations and assumptions, both explicit and hidden, any one of which could fail on a given combination of hardware, operating system, and language translator. For instance, the first two of the following conditions negate the potential gain in accuracy for every angle t, while the third affects the reduction only rather infrequently (Exercise 1a of Section 3.4):

1. If π_1 is specified in decimal, say, as 3.140625, then a binary value other than 11.001001 might be produced by the language translator.
2. A translator might apply a "constant folding optimization" to replace $(t - n\pi_1) - n\pi_2$ by $t - n\Pi$, where the translator computes $\Pi \leftarrow \pi_1 + \pi_2$.
3. The floating-point hardware might not provide a guard digit for subtraction.

Careful testing provides the only hope for verifying that our idealizations and assumptions are not violated in a harmful way.

Most straightforward approaches are inadequate for testing the correctness of high-accuracy sine procedures. For instance, existing sine procedures may not meet the exacting specification given below and hence cannot be used to generate values for purposes of comparison. Moreover, testing cannot be done using published tables of sine values unless the quoted arguments are exact machine numbers. Otherwise, converting an angle to a floating-point number could do more damage than all the errors committed by the sine procedure.

Sec. 3.5   Testing High-Accuracy Sine Procedures

A very simple accuracy test is to compare the computed value of $sin(22)$ with the exact value

$$sin(22) = -8.8513\,09290\,40388 \times 10^{-3}$$

As mentioned in the previous section, the argument $t = 22$ is particularly useful for testing sine procedures because four decimal digits cancel in the subtraction $22 - 7\pi$, and a procedure that uses a naive argument reduction scheme will suffer a corresponding loss of accuracy. Try this test on the next scientific calculator you see.

Systematic testing of a high-accuracy sine procedure requires that two hurdles be overcome. First, we need a specification to test. Second, we need a trustworthy alternative method to generate values of the sine function for purposes of comparison.

What Specification Can Be Tested?   Suppose that working precision is equivalent to using b-bit floating-point fractions and that the argument reduction step is done with the equivalent of $b + g$ bits. Thus if the value of π_1 is one of the two suggested in the previous section, then g is either 8 or 12. According to the analysis of Section 3.3, inaccuracy in $n\pi$ contributes an error of about $|t/sin(t)|$ ulps *of (b + g)-bit arithmetic*, which is equivalent to $2^{-g} \times |t/sin(t)|$ ulps in working precision, whereas the other errors (rounding x, inexact coefficients c_i, and so on) contribute roughly an additional ulp. We are thus led to formulate the following specification:

Specification 3.2
Let *g* be the number of extra bits of precision gained in the argument reduction step. The computed value of *sin*(*t*) should be correct to within

$$\beta \times \left(2^{-g} \times \left|\frac{t}{sin(t)}\right| + 1\right) \text{ ulps}$$

unless |*sin*(*t*) / *t*| is not much larger than ϵ.

What Can We Compare the Computed Value of *sin*(*t*) Against? Cody and Waite suggest a way to use the procedure being tested to generate values for comparison purposes. Their approach hinges on the "triple angle" identity

$$sin(t) = sin\left(\frac{t}{3}\right)\left[3 - 4\,sin^2\left(\frac{t}{3}\right)\right] \qquad (\dagger)$$

The idea is to sample a population of arguments t chosen so that

(C_1) $t/3$ can be computed exactly, and
(C_2) The error (in ulps) of the computed right-hand side of ($\dagger$) agrees closely with the error in the computed value of $sin(t/3)$.

If Specification 3.2 is met, then the computed left-hand side of (†) is within $\beta[2^{-g}|t/\sin(t)| + 1]$ ulps of the true value. If conditions C_1 and C_2 are also satisfied, then the computed right-hand side is correct to within $\beta[2^{-g}|(t/3)/\sin(t/3)| + 1]$ ulps. It follows that the computed sides of (†) should differ by at most the sum of these two amounts.

Thus, given an angle t satisfying conditions C_1 and C_2 and letting w be the number of ulps by which the computed sides of (†) differ, we can compute

$$\omega(t) = \frac{w}{2^{-g}|t/\sin(t)| + 2^{-g}|(t/3)/\sin(t/3)| + 2}$$

If, for some such t, $\omega(t)$ exceeds β, then Specification 3.2 must fail for either t or $t/3$. Without conditions C_1 and C_2, a large value of $\omega(t)$ would not necessarily signal a violation of Specification 3.2, since it might be caused by the use of inexact arithmetic to evaluate the right-hand side of (†). See Programming Assignment 2.

Meeting Condition C_1. Our approach is to make a minuscule change in every test angle t so as to guarantee that it has the form $t = 3u$, where u is a machine number. To do this we will perform the following "argument purification" procedure:

$$y \leftarrow \frac{t}{3}$$

$$u \leftarrow 5y - 4y$$

$$t \leftarrow 3 \times u$$

The only change in t comes as a result of rounding errors. The change can be at most a few ulps since the subtraction suffers only mild cancellation. To show that this argument purification procedure guarantees that condition C_1 will hold, we will show that the multiplication $t \leftarrow 3 \times u$ is exact.

First consider the case $\beta = 2$. Since the binary representation of 3 is 11, multiplication of u by 3 is equivalent to shifting u one place to the left (i.e., multiplying by 2) and adding the result to u. For instance,

```
      0.110101
  ×         11
      0.110101
      1.10101
     10.011111
```

The product $3 \times u$ can be represented in p bits (where p is the floating-point precision) if the last two bits of u's fraction are 0. This condition is guaranteed by the computation $u \leftarrow 5y - 4y$ since the floating-point exponent of $4y$ is two greater than that of u and hence two bits cancel in the subtraction. See Figure 3.10.

Sec. 3.5   Testing High-Accuracy Sine Procedures   69

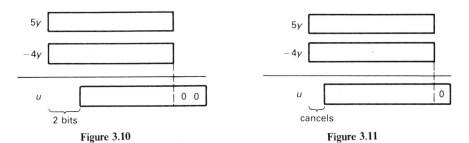

Figure 3.10   Figure 3.11

The reasoning for $\beta \neq 2$ is a little different. For instance, consider the case $\beta = 10$. Multiplication of a machine number

$$u = \pm 10^e \times f$$

where $f = 0.d_1 d_2 \cdots d_p$, by 3 is exact if either $f < \frac{1}{3}$ (so that multiplication by 3 does not involve a carry into the 1's position) or $d_p = 0$. If $f \geq \frac{1}{3}$, then $4y$ has a floating-point exponent $e + 1$, so one digit cancels in the subtraction $5y - 4y$, giving $d_p = 0$. See Figure 3.11. Thus in either case, i.e., $f < \frac{1}{3}$ or $f \geq \frac{1}{3}$, $3 \times x$ will be computed without error. For other bases $\beta > 3$ the reasoning is essentially identical, while in the unlikely case that $\beta = 3$, multiplication by 3 is always exact.

A warning is in order. If the computer evaluates $5y$, $4y$, and $5y - 4y$ with extra precision before converting the result to working precision for assignment to u, then u's trailing bits will not be set to 0. (Recall the discussion in Section 2.2 about the pitfalls of computing values of environmental parameters.) This can be circumvented by forcing the computer to store intermediate values, as with $y5 \leftarrow 5 \times y$, $y4 \leftarrow 4 \times y$, $u \leftarrow y5 - y4$.

Meeting Condition C_2. The accuracy of $sin(t/3)[3 - 4 sin^2(t/3)]$ will match that of $sin(t/3)$ provided that the accuracy of $3 - 4 sin^2(t/3)$ equals or exceeds that of $sin(t/3)$. (Recall, from Section 2.4, that the accuracy of a product roughly equals that of its least accurate factor.) For any number s, the accuracy of $4s^2$ is very nearly that of s, so the problem reduces to one of eliminating the possibility of cancellation in the subtraction operation. This can be accomplished by testing only with arguments of the form $t = 3m\pi + z$, where m is an integer and $|z| \leq \pi/2$. Doing so guarantees that $|t/3 - m\pi| \leq \pi/6$, so that $|sin(t/3)| \leq \frac{1}{2}$, $|4 sin^2(t/3)| \leq 1$, and there can be no appreciable cancellation in the subtraction $3 - 4 sin^2(t/3)$. In fact, since $3 - 4 sin^2(t/3)$ is at least twice the size of $4 sin^2(t/3)$, one or more inaccurate bits will be discarded when the floating-point fraction of $4 sin^2(t/3)$ is shifted to the right for the subtraction. See Figure 3.12. With this choice, the arguments t still cover all possible values x obtained by argument reduction, thereby thoroughly exercising the basic approximation scheme.

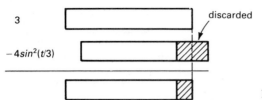

Figure 3.12

PROGRAMMING ASSIGNMENTS

1. Test your high-precision sine procedure using the strategy developed in this section. [*Warning*: Not only is multiplication by the expression $2**-8$ an inefficient and inaccurate way of dividing by 256, but some systems evaluate the expression $2**-8$ as 0.]
2. (Optional) Try testing your high-precision sine procedure using test angles t that fail to satisfy one of the above conditions C_1 or C_2. Does $\omega(t)$ appear to exceed β?

EXERCISES

1. Discuss the usefulness of the identity $sin^2(t) + cos^2(t) = 1$ for testing sine and cosine procedures. [*Hint*: Consider a sine-cosine procedure in which the argument reduction step returns $x = 0.0$ for any argument t.]
2. Show that the above testing strategy is inadequate to expose the failure of the argument reduction step $x \leftarrow (t - n\pi_1) - n\pi_2$ on a binary ($\beta = 2$) machine that lacks a guard digit for subtraction. [*Hint*: Recall Exercise 1 of Section 3.4. No problem can arise in evaluating $sin(t/3)$ since the fraction portion of $u = t/3$ ends with a zero. The only possibility of failure is when t lies just under some 2^k. But in that case, the exponent of t is only one larger than that of $t/3$, and the last digit of t must be zero.]
3. (Optional) What precautions are necessary if the identity

$$cos(t) = cos\left(\frac{t}{3}\right)\left[4 cos^2\left(\frac{t}{3}\right) - 3\right]$$

is used to test the accuracy of a cosine procedure?
4. (Optional and difficult) Consider the following argument purification procedure:

$$y \leftarrow \frac{t}{3}$$
$$u \leftarrow (y + t) - t$$
$$t \leftarrow 3 \times u$$

(a) Extend the argument given in this section to "prove" that the multiplication $3 \times u$ is exact. Note explicitly any uses of the assumption $t = 3u$ where t

denotes the *initial* value. [*Hint*: If $\beta = 2$ and u's fraction is less than $\frac{2}{3}$, then only one bit cancels in the subtraction. However, that is adequate since the floating-point exponent of $3u$ is only 1 larger than u's.]

(b) Imagine a machine with $\beta = 2$ and $p = 3$. Show that the following values are computed from an initial value $t = 0.111$: $y \leftarrow 0.0101$ [*Hint*: Use the shift-and-add technique to find $3 \times 0.100101010\ldots$], $u \leftarrow 0.011$, and $t \leftarrow$ either 1.00 or 1.01. Can the resulting value of t be divided by 3 without error?

(c) Show that the result of Exercise 2 also holds for this argument purification procedure.

4

LINEAR EQUATIONS

It is natural for matrix programs, especially programs to solve linear equations, to play a central role in any study of numerical software. These programs constitute the most important class of numerical software for which there exist specifications, that is, precise and complete requirements for what a program should compute. Moreover, they have long served as a focus for research on engineering techniques for the production of numerical software.

The main thrust of our study of linear equations will be divided about equally between implementation techniques and the question, *Is testing a reliable method for deciding whether a program meets a specification?* In addition, a use of performance measurements will be noted.

We begin with a review of basic material about systems of n linear equations in n unknowns. This problem can be compactly posed in terms of matrices and vectors as follows:

The Problem of n Linear Equations in n Unknowns
Let A be a matrix with n rows and n columns and let **b** be a vector with n entries. Find a vector **x** with n entries that satisfies $A\mathbf{x} = \mathbf{b}$.

Section 4.1 presents several linear equation solvers, that is, algorithms to solve the problem of n linear equations in n unknowns. Most of the section proceeds quite briskly because we assume that the reader has an acquaintance with the material.

Section 4.2 discusses various issues that arise in the process of turning matrix algorithms into reliable and efficient computer programs. In an attempt to gain efficiency, we are led to consider such diverse matters as computer architecture and compiler code-generation details. However, the only way to be certain that efficiency has been improved is to measure the program's performance. Much of this material about implementation has been extracted from the *LINPACK Users' Guide* by J. J. Dongarra, J. R. Bunch, C. B. Moler, and G. W. Stewart (SIAM, Philadelphia, 1979), which describes an extremely high-quality collection of matrix software.

Theoretical analyses of rounding error propagation in a matrix algorithm (i.e., proofs that the algorithm satisfies a certain specification) are typically so difficult that only a specialist in matrix computations can be expected to complete one. No such analysis will be given in this book; the interested reader can consult *Computer Solution of Linear Algebraic Systems* by George Forsythe and Cleve Moler (Prentice-Hall, Englewood Cliffs, N.J., 1967) or James Wilkinson's books: *Rounding Errors in Algebraic Processes* (Prentice-Hall, Englewood Cliffs, N.J., 1963) and *The Algebraic Eigenvalue Problem* (Oxford University Press, London, 1965). Even experts often find these theoretical analyses repugnant. The following comments of Beresford Parlett in *The Symmetric Eigenvalue Problem* (Prentice-Hall, Englewood Cliffs, N.J., 1980, p. xv) summarize the feelings of many:

> My heart always sinks when the subject of rounding errors is mentioned. The sometimes bizarre effects of fixed precision arithmetic should be at the heart of work in matrix computations. Yet formal treatment of roundoff, though it seems necessary, rarely enlightens me.

In Section 4.3 we show that careful testing procedures can remove some of the need for the theoretical analyses so distasteful to Parlett, though they are not a complete substitute. The algorithmic failures that are hard to detect are the "numerical instabilities" associated with fixed-precision arithmetic. Many of the numerically unstable linear equation solvers that arise in practice can be detected by a well-designed test using a large number of systems of, say, 4 linear equations in 4 unknowns. However, a few numerically unstable procedures are quite hard to detect by such tests because either (1) they fail too rarely on 4×4 problems or (2) they fail only for larger systems of equations, say, 30 equations in 30 unknowns.

Most experts prefer Gaussian elimination with partial pivoting for solving linear equations, unless the matrix A possesses some special property. As discussed in Section 4.4, a major part of the justification for this choice rests on performance measurements showing that the computed results "almost always" satisfy certain conditions.

No discussion of linear equation solvers can be considered complete without mention of the accuracy of the computed solution. Interesting software issues arise when one tries to provide the user with an error estimation (or what is essentially equivalent, with an estimation of the "condition number" of his problem) that is both economical and trustworthy. For instance, testing can be far less reliable for exposing errors in condition estimators than it is for errors in other parts of linear equation solvers. Unfortunately, we have had to omit this subject because it is substantially harder to understand than other topics discussed in this book. Most of the same general points about numerical software will be observed in later chapters.

4.1 ELEMENTARY FACTS ABOUT LINEAR EQUATIONS

In this section we will review some basic facts about linear equations and discuss methods for solving them. We begin with two formulations of the problem.

Formulation 1. Given numbers a_{ij} and b_i, $1 \le i \le n$ and $1 \le j \le n$, find numbers $x_1, x_2, \ldots, x_n$ such that

$$a_{11}x_1 + a_{12}x_2 + \cdots + a_{1n}x_n = b_1$$
$$a_{21}x_1 + a_{22}x_2 + \cdots + a_{2n}x_n = b_2$$
$$\vdots$$
$$a_{n1}x_1 + a_{n2}x_2 + \cdots + a_{nn}x_n = b_n$$

Sometimes it is desirable to think of the coefficients a_{ij} as forming an $n \times n$ matrix A, that is, a square array of numbers having n rows and n columns:

$$A = \begin{bmatrix} a_{11} & a_{12} & \cdots & a_{1n} \\ a_{21} & a_{22} & \cdots & a_{2n} \\ \vdots & \vdots & \vdots & \vdots \\ a_{n1} & a_{n2} & \cdots & a_{nn} \end{bmatrix}$$

The solution can be thought of as an n-vector

$$\mathbf{x} = \begin{bmatrix} x_1 \\ x_2 \\ \vdots \\ x_n \end{bmatrix}$$

We will write the i, j-entry of A, that is, the entry lying at the intersection of the ith row and the jth column, as a_{ij}. The product $A\mathbf{x}$ of an $n \times n$ matrix A and an n-vector $\mathbf{x}$ is defined to be the n-vector with i^{th} entry $\sum_{k=1}^{n} a_{ik}x_k$. In other words, the ith entry of $A\mathbf{x}$ is the left-hand side of the ith equation in

Sec. 4.1 Elementary Facts About Linear Equations

Formulation 1. This definition leads immediately to the following short-hand notation for the linear equation problem.

Formulation 2. Given an $n \times n$ matrix A and an n-vector **b**, find an n-vector **x** such that $A\mathbf{x} = \mathbf{b}$.

Depending on the particular coefficients a_{ij} and b_i, a set of linear equations may have no solution, exactly one solution, or infinitely many solutions. In terms of computation, this means that the algorithms given below could, in theory, call for division by zero. Thus the algorithms should be implemented so as to report failure in the rare circumstances that the solution process is not defined.

A set of linear equations is said to be *ill-conditioned* if, by slight alterations to its coefficients, it can be turned into a set of linear equations with no solution. Ill-conditioned linear equations suffer from two ailments:

1. Even tiny errors in the data can drastically change the solution.
2. Numerical solution of the equations, even by the best available method, will almost certainly produce an inaccurate solution.

For example, the linear equations $A\mathbf{x} = \mathbf{b}$, where

$$A = \begin{bmatrix} 0.578 & 0.323 \\ 0.377 & 0.212 \end{bmatrix}, \quad \mathbf{b} = \begin{bmatrix} 0.901 \\ 0.589 \end{bmatrix}$$

are ill-conditioned. Whereas the true solution is

$$\mathbf{x} = \begin{bmatrix} 1.0 \\ 1.0 \end{bmatrix}$$

changing the 2,2-entry of A from 0.212 to 0.211 changes the solution to

$$\mathbf{x} = \begin{bmatrix} -0.72\ldots \\ 4.09\ldots \end{bmatrix}$$

an unmistakable symptom of ailment 1. Also, the solution computed in three-decimal arithmetic by the most accurate method discussed below is

$$\mathbf{x} = \begin{bmatrix} 0.441 \\ 2.00 \end{bmatrix}$$

illustrating ailment 2.

Notice the similarities between the problem of solving an ill-conditioned set of linear equations and the problem of computing $sin(t)$ where t is close to a multiple of π. For such angles t a one-ulp error in t changes the solution $sin(t)$ by many [roughly, $|t/sin(t)|$] ulps, and rounding errors in the argument reduction step will almost certainly lead to an inaccurate computed result.

Gaussian Elimination. Let us work through a simple example before we consider a general formulation of the algorithm. Consider the equations

$$4x_1 + 3x_2 + 0x_3 = 7.5$$
$$2x_1 - 1x_2 + 1x_3 = 3.5$$
$$1x_1 + 0x_2 + 2x_3 = 3.5$$

If we subtract 0.5 times the first equation from the second, and subtract 0.25 times the first equation from the third, we get the new equations:

$$4x_1 + 3x_2 + 0x_3 = 7.5$$
$$-2.5x_2 + 1x_3 = -0.25$$
$$-0.75x_2 + 2x_3 = 1.625$$

Subtracting 0.3 times the modified second equation from the third gives the following "triangular" equations:

$$4x_1 + 3x_2 + 0x_3 = 7.5$$
$$-2.5x_2 + 1x_3 = -0.25$$
$$1.7x_3 = 1.7$$

It follows immediately from the last equation that $x_3 = 1$. Substituting this value in the second equation gives $-2.5x_2 + 1 = -0.25$, which simplifies to $-2.5x_2 = -1.25$ or $x_2 = 0.5$. Finally, substituting these values of x_2 and x_3 in the first equation and solving for x_1 gives $x_1 = 1.5$.

We are now ready to tackle the general problem. The set of solutions of a system of linear equations is preserved if, for some k and i with $1 \leq k \leq n$ and $1 \leq i \leq n$, a constant multiple of the kth equation is subtracted from the ith equation (with the result replacing the ith equation). If $a_{11} \neq 0$, then by

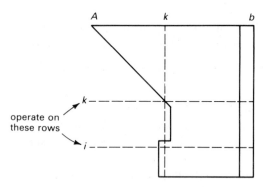

Figure 4.1

Sec. 4.1 Elementary Facts About Linear Equations

subtracting a certain multiple *amult* of the first equation we can "reduce a_{i1} to zero," that is, arrange for the coefficient of x_1 in the resulting i^{th} equation to be zero. In particular, *amult* must satisfy $a_{i1} - amult \times a_{11} = 0$, so *amult* = a_{i1}/a_{11}.

In general, suppose we have performed these row operations so as to reduce to zero all coefficients below the diagonal in columns 1 through $k - 1$ and all coefficients in column k that lie in rows $k + 1$ through $i - 1$. See Figure 4.1. Then a_{ik} is reduced to zero by subtracting *amult* = a_{ik}/a_{kk} times equation k from equation i. (Here a_{ik} and a_{kk} denote values that may have been produced by earlier row operations.)

Completion of this process results in an upper triangular set of equations; see Figure 4.2. The last equation is now $a_{nn}x_n = b_n$, which immediately gives x_n. Suppose that $x_n, x_{n-1}, \ldots, x_{i+1}$ have been found. Equation i has the form

$$a_{ii}x_i + a_{i,i+1}x_{i+1} + \cdots + a_{in}x_n = b_i$$

or $a_{ii}x_i + \sum_{j=i+1}^{n} a_{ij}x_j = b_i$, which can be solved for x_i if $a_{ii} \neq 0$.

The entire Gaussian elimination process is given below as Algorithm 4.1a. One aspect of our notation for expressing matrix algorithms, namely the meaning of degenerate cases of *for* loops and sum operators, deserves an explanation. If the lower limit of a *for* loop is larger than the upper limit, then the loop is not executed. For example, if $k = n$, then the body of the loop

$$\text{for } i = k + 1 \text{ to } n$$

is not executed. [*Warning*: With many old FORTRAN compilers, *DO* loops are always executed at least once, so care must be taken when implementing *for* statements.] Similarly, a sum is defined to equal zero if its lower limit exceeds its upper limit. Thus if $i = n$, then $(b_i - \sum_{j=i+1}^{n} a_{ij} \times x_j)/a_{ii}$ equals b_i/a_{ii}.

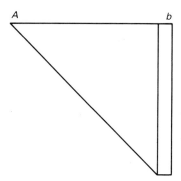

Figure 4.2

Algorithm 4.1a
..Gaussian elimination.
..Reduce to upper triangular equations.

for $k = 1$ to $n - 1$

    for $i = k + 1$ to n

        amult $\leftarrow a_{ik} / a_{kk}$

        for $j = k + 1$ to n

            $a_{ij} \leftarrow a_{ij} - $ amult $\times a_{kj}$

        $b_i \leftarrow b_i - $ amult $\times b_k$

..Solve the resulting upper triangular equations.

for $i = n$ down to 1

    $x_i \leftarrow (b_i - \sum_{j=i+1}^{n} a_{ij} \times x_j) / a_{ii}$

Unfortunately this algorithm is numerically unstable (in a sense to be made precise in Section 4.3). A thorough analysis by James Wilkinson has shown that the instability is intimately connected with growth in the sizes of the coefficients a_{ij} during the triangularization step. One strategy for reducing this growth, called "partial pivoting," involves interchanging rows so as to guarantee that no multiplier *amult* exceeds 1 in absolute value. In particular, just before zeros are introduced in column k, row k can be interchanged with a later row to bring the largest of $a_{kk}, a_{k+1,k}, \ldots, a_{nk}$ up into the kk position.

The resulting procedure is shown in Algorithm 4.1b. (The notation "$s \leftrightarrow t$" means that the values of s and t are to be interchanged.)

Operation Counts. Rough comparisons of algorithm execution speeds can be made by counting the numbers of floating-point operations. For most purposes it is sufficient to calculate only the leading term of the polynomial $p(n)$ that gives the number of floating-point operations in a matrix algorithm as a function of the data size. This allows us to ignore all operations except those in the deepest loops and to make approximations like

$$\sum_{i=1}^{n} i = \frac{n(n+1)}{2} \approx \frac{n^2}{2}$$

$$\sum_{i=1}^{n} i^2 = \frac{n(n+1)(2n+1)}{6} \approx \frac{n^3}{3}$$

and

$$\sum_{i=1}^{n}(n-i) = (n-1) + (n-2) + \cdots + 1 + 0 = \sum_{i=1}^{n-1} i \approx \frac{n^2}{2}$$

Sec. 4.1 Elementary Facts About Linear Equations

Algorithm 4.1b
..Gaussian elimination with partial pivoting.
..Reduce to upper triangular equations.
for $k = 1$ to $n - 1$
..Find the largest subdiagonal element in column k.
$m \leftarrow k$
for $i = k + 1$ to n
if $|a_{ik}| > |a_{mk}|$
$m \leftarrow i$
..Interchange rows m and k.
for $j = k$ to n
$a_{mj} \leftrightarrow a_{kj}$
$b_m \leftrightarrow b_k$
..Continue as before.
for $i = k + 1$ to n
amult $\leftarrow a_{ik} / a_{kk}$
for $j = k + 1$ to n
$a_{ij} \leftarrow a_{ij} - $ amult $\times a_{kj}$
$b_i \leftarrow b_i - $ amult $\times b_k$
..Solve the resulting upper triangular equations.
for $i = n$ down to 1
$$x_i \leftarrow \left(b_i - \sum_{j=i+1}^{n} a_{ij} \times x_j \right) / a_{ii}$$

Even with these simplifications, counting operations requires a fair amount of concentration.

Example 1

For the purposes of counting multiplications, the algorithm that solves an upper triangular system $U\mathbf{x} = \mathbf{c}$

for $i = n$ down to 1
$$x_i \leftarrow \left(c_i - \sum_{j=i+1}^{n} u_{ij} \times x_j \right) \Big/ r_{ii}$$

can be perceived as

> for $i = 1$ to n
>
> > $n - i$ multiplications

Thus the total number of multiplications is $\sum_{i=1}^{n}(n - i) \approx n^2/2$. For this algorithm, as for almost all matrix algorithms, the number of additions and subtractions is essentially the same as the number of multiplications.

Example 2

Gaussian elimination, stripped of operations that cannot appreciably affect a count of the floating-point multiplications, looks like this:

> for $k = 1$ to n
>
> > for $i = k + 1$ to n
> >
> > > for $j = k + 1$ to n
> > >
> > > > 1 addition and 1 multiplication

or, equivalently,

> for $k = 1$ to n
>
> > for $i = k + 1$ to n
> >
> > > $n - k$ additions and $n - k$ multiplications

The total number of floating-point multiplications is

$$\sum_{k=1}^{n} \sum_{i=k+1}^{n} (n - k) = \sum_{k=1}^{n} (n - k)^2$$
$$= (n - 1)^2 + \cdots + 2^2 + 1^2$$
$$= \sum_{k=1}^{n-1} k^2$$
$$\approx \frac{n^3}{3}$$

EXERCISES

1. Using words and pictures, argue that Algorithm 4.1c computes a vector **x** satisfying $A\mathbf{x} = \mathbf{b}$. (Neglect floating-point errors and assume that all operations are defined.)

2. Count the floating-point operations in Algorithm 4.1c, and compare the number with the count for Gaussian elimination.

4.2 IMPLEMENTATION DETAILS

In this section we will address the following five implementation issues: (1) proper packaging, (2) time efficiency, (3) storage efficiency, (4) overflow and underflow, and (5) program variants. For implementing Gaussian elimination,

Algorithm 4.1c
..Gauss-Jordan elimination with partial pivoting.

..Reduce to a diagonal system of equations.

   for $k = 1$ to n

      ..Find the largest subdiagonal element in column k.

      $m \leftarrow k$

      for $i = k + 1$ to n

         if $|a_{ik}| > |a_{mk}|$

            $m \leftarrow i$

      ..Interchange rows m and k.

      for $j = k$ to n

         $a_{mj} \leftrightarrow a_{kj}$

      $b_m \leftrightarrow b_k$

      ..Reduce to zero all nondiagonal elements in column k.

      for $i = 1$ to n

         if $i \neq k$

            amult $\leftarrow a_{ik} / a_{kk}$

            for $j = k + 1$ to n

                $a_{ij} \leftarrow a_{ij} -$ amult $\times a_{kj}$

            $b_i \leftarrow b_i -$ amult $\times b_k$

..Solve the resulting diagonal equations.

   for $i = 1$ to n

      $x_i \leftarrow b_i / a_{ii}$

items 1, 2, 3, and 5 are important. In fact, the approach advocated below for dealing with program variants is completely general, rather than being specific to matrix software. Item 4 is discussed briefly because it is crucial for many other matrix algorithms.

4.2.1 Proper Packaging

It is desirable to implement a linear equation solver as two procedures. Such a division of duties makes possible the economical solution of a sequence of linear equations with the same matrix A and different vectors $\mathbf{b}_1, \mathbf{b}_2, \ldots$, even if $\mathbf{b}_2$ cannot be determined until the problem is solved for $\mathbf{b}_1$. (This application arises with the "iterative improvement" of solutions $\mathbf{x}$.)

The first procedure reduces A to an upper triangular matrix U and makes a record of the operations that are to be performed on $\mathbf{b}$. (It may be possible to use memory locations corresponding to elements of A that have been reduced to zero to hold information about the transformation that did the reduction.) The second procedure applies the recorded operations to $\mathbf{b}$ and then solves the resulting system $U\mathbf{x} = \mathbf{c}$. For a FORTRAN implementation, the following design might be used. (In practice, calling sequences may well turn out to be longer than these, but care should be taken to avoid unnecessary proliferation of subprogram arguments.)

1. SUBROUTINE REDUCE (A, LDA, N, IFAIL). This subroutine reduces to upper triangular form an $N \times N$ matrix that occupies a storage area with leading dimension LDA. If *IFAIL* has a nonzero value upon return from *REDUCE*, then the reduced matrix has a zero on its diagonal so an attempt to solve linear equations involving A would result in division by zero. Otherwise *IFAIL* is set to 0 and the values returned in A can be used to solve linear equations.

2. SUBROUTINE SOLVE (A, LDA, N, B, X). *SOLVE* can be applied to a matrix that has been reduced using *SUBROUTINE REDUCE* (provided that *IFAIL* = 0 was returned) and to an N-vector B. The solution of the linear equations is returned in X.

The argument *LDA* is peculiar to FORTRAN implementations, where subprograms need to know the declared number of rows, e.g., the *50* in the calling-program declaration *REAL A(50, 20)*, but not the declared number of columns. The reason is that the declared number of rows of A, or, equivalently, the declared length of a column, determines the spacing in computer memory between successive entries along a row of A. (See Exercise 1.) For instance, if the calling program declares

```
REAL A(50,50), B(50), X(50)
```

and solves a 20 × 20 problem by

```
CALL REDUCE(A, 50, 20, IFAIL)
IF (IFAIL .NE. 0) GO TO 1000
CALL SOLVE(A, 50, 20, B, X)
```

then *SOLVE* can declare

```
REAL A(LDA,N), B(N), X(N)
```

even if the compiler generates code that checks at execution time whether array subscripts are in bounds.

4.2.2 Time Efficiency

Efficiency is important for linear equation solvers. A typical use of multi-million-dollar supercomputers is to perform computations that reduce to the solution of many sets of linear equations. Thus doubling the speed of a linear equation solver might halve the number of computers that are needed for such work.

In many circumstances, less computation time is spent doing floating-point operations than is spent moving data and computing addresses in computer memory. Techniques for reducing the time required for these other operations are discussed next. The success of such techniques is highly computer- and compiler-dependent [see "The Influence of the Compiler on the Cost of Mathematical Software" by B. N. Parlett and Y. Wang (*ACM Transactions on Mathematical Software*, March 1975, pp. 35–46)]. Any attempt to strike a balance that works reasonably well for many machines can be evaluated only by multimachine performance measurements.

The LINPACK collection of matrix programs achieves a times savings of as much as a factor of 2 on some machines by the use of a few simple programming conventions for inner loops. One technique, specific to the way that FORTRAN stores matrices, decreases the movement of data that many computers perform automatically. Another technique, applicable to most programming languages, decreases the overhead of loop indexing. Both techniques, while degrading performance on a few machines (particularly for small problems), enhance performance on others for a variety of reasons.

Column Orientation. LINPACK adopts the convention that inner loops access matrix entries by columns. FORTRAN stores matrices by columns, so this approach generates sequential accesses to memory, whereas writing inner loops to march across rows causes successive memory references to be widely spaced in memory. This column orientation substantially improves performance on many machines, particularly those that utilize cache memory or paged virtual memory.

It takes some concentration to see that the following procedure performs exactly the same floating-point operations as does Algorithm 4.1b to reduce the original equations to triangular form, though the operations are not done in the same order. Solution of the triangular equations is not done in quite the same way, however. Instead of adding the terms $a_{ij}x_j$ for $j = i + 1$ to n and then subtracting the sum from b_i, the terms are subtracted from b_i one at a time. [A general theorem in "Reducibility Among Floating-Point Graphs" by Donald Johnson, Webb Miller, Brian Minnihan and Celia Wrathall (*Journal of ACM*, Oct. 1979, pp. 739–760, especially p. 753) shows that this sort of transformation to the algorithm cannot affect its numerical stability or instability.]

Algorithm 4.2
..Gaussian elimination with partial pivoting.
..(Column-oriented version.)

..Reduce to upper triangular equations.

    for $k = 1$ to $n - 1$

        ..Find the largest subdiagonal element in column k.

        $m \leftarrow k$

        for $i = k + 1$ to n

            if $|a_{ik}| > |a_{mk}|$

                $m \leftarrow i$

        ..Move the pivot element to the diagonal.

        $a_{mk} \leftrightarrow a_{kk}$

        ..Overwrite with the multipliers.

        for $i = k + 1$ to n

            $a_{ik} \leftarrow a_{ik} / a_{kk}$

        for $j = k + 1$ to n

            ..Interchange rows...

            $a_{mj} \leftrightarrow a_{kj}$

            ..and perform the row operation.

            for $i = k + 1$ to n

                $a_{ij} \leftarrow a_{ij} - a_{ik} \times a_{kj}$

        ..Interchange entries of **b**

        $b_m \leftrightarrow b_k$

        for $i = k + 1$ to n

            $b_i \leftarrow b_i - a_{ik} \times b_k$

..Solve the resulting upper triangular equations.

    for $j = n$ down to 1

        $x_j \leftarrow b_j / a_{jj}$

        for $i = 1$ to $j - 1$

            $b_i \leftarrow b_i - a_{ij} \times x_j$

For a discussion of other column-oriented variants of Gaussian elimination, see "Solving Large Full Sets of Linear Equations in a Paged Virtual Store" by J. J. DuCroz, S. M. Nugent, J. K. Reid, and D. B. Taylor (*ACM Transactions on Mathematical Software*, Dec. 1981, pp. 527–536).

Unrolled Loops. In LINPACK, inner loops computing $\sum_i x_i y_i$ are performed by calling a function like the following. (We have changed the names and some of the details of the following two subprograms.)

```
      REAL FUNCTION SPROD(X,Y,LOWER,UPPER)
      REAL X(*), Y(*)
      INTEGER LOWER, UPPER, M, I
      SPROD = 0.0
      M = MOD(UPPER - LOWER + 1, 5)
      DO 10 I = LOWER, LOWER + M - 1
         SPROD = SPROD + X(I)*Y(I)
   10 CONTINUE
      DO 20 I = LOWER + M, UPPER, 5
         SPROD = SPROD + X(I)*Y(I) + X(I+1)*Y(I+1) +
     *      X(I+2)*Y(I+2) + X(I+3)*Y(I+3) +
     *      X(I+4)*Y(I+4)
   20 CONTINUE
      RETURN
      END
```

Similarly, the following subroutine performs the vector operation $y \leftarrow \alpha x + y$:

```
      SUBROUTINE SADD(ALFA,X,Y,LOWER,UPPER)
      REAL ALFA, X(*), Y(*)
      INTEGER LOWER, UPPER, M, I
      M = MOD(UPPER - LOWER + 1, 4)
      DO 10 I = LOWER, LOWER + M - 1
         Y(I) = ALFA*X(I) + Y(I)
   10 CONTINUE
      DO 20 I = LOWER + M, UPPER, 4
         Y(I) = ALFA*X(I) + Y(I)
         Y(I+1) = ALFA*X(I+1) + Y(I+1)
         Y(I+2) = ALFA*X(I+2) + Y(I+2)
         Y(I+3) = ALFA*X(I+3) + Y(I+3)
   20 CONTINUE
      RETURN
      END
```

For example the inner loop

$$\text{for } i = k + 1 \text{ to } n$$
$$a_{ij} \leftarrow a_{ij} - a_{ik} \times a_{kj}$$

of Algorithm 4.2 can be implemented in single precision as

```
CALL SADD( -A(K,J), A(1,K), A(1,J), K+1, N)
```

SADD interprets its second and third arguments, which in this case are the addresses in memory of the first entries in columns K and J of the matrix A, as the locations of the first entries in two sequences of consecutive locations of type *real*. Thus a reference in *SADD* to $X(I)$ names the $I - 1^{st}$ location after $A(1, K)$, i.e., $A(I, K)$.

Extensive measuring on a variety of computer-compiler combinations has shown the two loops to be generally efficient. The reasons why they save execution time and the amounts of time saved vary widely, but the general idea is to avoid 80% (for *SPROD*) or 75% (for *SADD*) of the cleanup operations that are performed at the foot of a *DO* loop. On one machine, the first program segment runs slightly slower than the usual loop for computing inner products, while on another it runs about 145% faster. The numbers of terms in the inner loop, 5 for *SPROD* and 4 for *SADD*, were chosen because raising them makes the programs run slower on one economically important family of computers. For a thorough discussion, see "Unrolling Loops in FORTRAN" by J. J. Dongarra and A. R. Hinds (*Software—Practice and Experience*, March 1979, pp. 219–226).

For certain machines, the use of subprograms to perform inner-loop computations provides additional advantages. Many compilers produce more efficient machine code for singly subscripted arrays than for doubly subscripted arrays [e.g., $X(I)$ instead of $A(I, J)$]. Moreover, writing machine language versions of these subprograms may be an inexpensive way of saving a considerable amount of machine time, particularly on some of the newer computers with exotic architectures. For information about an available library of FORTRAN subprograms for inner loops of matrix computations, which includes the procedures used by LINPACK, see "Basic Linear Algebra Subprograms for FORTRAN Usage" by C. L. Lawson, R. J. Hanson, D. R. Kincaid, and F. T. Krogh (*ACM Transactions on Mathematical Software*, Sept. 1979, pp. 308–323).

4.2.3 Storage Efficiency

In some contexts a program's storage requirements are vitally important; other times they matter very little. Currently we see both an explosive growth in the use of microcomputers with limited amounts of memory and a trend toward "virtual memory" in large computers, so it is hard to say whether storage efficiency is increasing or decreasing in importance. In any case, there are several more-or-less obvious means of conserving storage in matrix computations.

Overwriting Storage. Storage may be recycled when it holds values that are no longer needed. The goal is to come as close as possible to using no array storage beyond that required to hold the data. For instance, the *REDUCE* procedure can use the locations in *A* that lie below the diagonal to pass the multipliers on to *SOLVE*.

Packing Storage. Instead of wasting space by using square arrays to hold triangular or symmetric matrices, one can pack their entries into singly subscripted arrays. Devising such implementations is not conceptually deep, just tedious. The interested reader might begin by studying the LINPACK codes to factor a symmetric, positive definite matrix; see especially the listings on pp. C.38–C.41 of the *LINPACK Users' Guide*.

Sparsity. Many computational problems that arise in practice involve a matrix that is large and *sparse*, i.e., that contains so many zeros that it is worthwhile to organize the computation so as to avoid storing and computing with zeros. Algorithms that transform such matrices must be evaluated on the basis of how well they preserve sparsity. The effort required to build an adequate piece of software is far greater for large, sparse matrix problems than it is for small, "dense" problems, and, likewise, a detailed discussion of sparse matrix software is beyond the scope of this book. The interested reader is referred to *Computer Solution of Large Sparse Positive Definite Systems* by Alan George and Joseph Liu (Prentice-Hall, Englewood Cliffs, N.J., 1981).

4.2.4 Overflow and Underflow

If it is possible to do so without incurring a significant penalty in execution costs, a matrix procedure should be written so as to lessen the likelihood of overflow or underflow. With Gaussian elimination, there is little that can be done along these lines without significantly degrading performance, but for many related matrix procedures there exist cost-effective precautions.

In particular, many matrix algorithms expend a small proportion of their floating-point operations in computations of the form

$$z \leftarrow \left(\sum_i v_i^2 \right)^{\frac{1}{2}}$$

For about half of all machine numbers v, v^2 either overflows or underflows, so direct application of this expression for z may well result in arithmetic anomalies. However, we can instead compute

$$t \leftarrow max(|v_i|)$$

$$z \leftarrow t \times \left[\sum_i \left(\frac{v_i}{t} \right)^2 \right]^{\frac{1}{2}}$$

The sum is of a sequence of nonnegative numbers, the largest of which is 1. Overflow is impossible unless **v** has an astronomical number of entries. Furthermore, adding an underflowed value (which we assume is set to zero) to a sum that is larger than 1 can do no harm. Of course t can be so large that z overflows, but in general that case cannot be salvaged by any reasonable means.

4.2.5 Program Variants

All major producers of numerical software libraries make use of the computer to help control the proliferation of programs that must be managed. Even if every effort is made to enhance portability, it is still necessary to produce several versions of a program. Rather than saving all variants of a program, it is sufficient to keep a "master" copy from which the variants can be generated automatically.

The following example is modeled on the use of the *M4* macro processor and of the compiler for *EFL* (short for *E*xtended *F*ortran *L*anguage), both of which are distributed with the UNIX operating system. (See *The Bell System Technical Journal*, July–Aug. 1978.) While these software tools were not specifically designed to assist maintenance of numerical software libraries, they can be used to illustrate some of the capabilities of the more specialized systems.

Suppose we need a subroutine that solves linear equations $T\mathbf{x} = \mathbf{b}$, where T is lower triangular, using the following column-oriented algorithm:

for $j = 1$ to n

$x_j \leftarrow b_j / t_{jj}$

for $i = j + 1, n$

$b_i \leftarrow b_i - t_{ij} \times x_j$

(The ith equation of $T\mathbf{x} = \mathbf{b}$ reads

$$t_{i1}x_1 + t_{i2}x_2 + \cdots + t_{ii}x_i = b_i$$

and this procedure computes x_i by subtracting $t_{ij}x_j$ from b_i for $j = 1, 2, \ldots, i - 1$ and then dividing the result by t_{ii}. This is essentially the same idea as is used in the back-substitution portion of Algorithm 4.2.) Suppose further that

Sec. 4.2   Implementation Details

we need to have the following variants of the procedure:

Version 1: single precision; inner loop written as a *DO* loop

Version 2: double precision; inner loop written as a *DO* loop

Version 3: single precision; inner loop written as a call to *SADD* (see above)

Version 4: double precision; inner loop written as a call to *DADD*, the double precision version of *SADD*

Our master version of the program follows. Most of it is copied directly from a more general procedure in LINPACK. We have used the convention of capitalizing those portions of the program text that depend on which version is desired.

```
            subroutine trisol(t, ldt, n, b, ifail)
            integer ldt, n, ifail
            TYPE t(ldt, *), b(*)
*
*           trisol solves lower triangular linear systems of
*           equations.
*
*           on entry
*
*                   t       TYPE array
*                           t contains the matrix of the system.
*                           The zero elements of the matrix are
*                           not referenced, and the corresponding
*                           elements of the array can be used
*                           to store other information.
*
*                   ldt     integer
*                           ldt is the leading dimension of the
*                           array t.
*
*                   n       integer
*                           n is the order of the system.
*
*                   b       TYPE array
*                           b contains the right-hand side of the
*                           system.
*
*           on return
*
*                   b       contains the solution, if ifail .eq. 0.
*                           otherwise b is unaltered.
*
```

```
*              ifail   integer
*                      ifail contains zero if the system is
*                      nonsingular. Otherwise ifail contains
*                      the subscript of the first zero
*                      diagonal element of t.
*
*      internal variables.
       integer j
       TYPE temp

       do ifail = 1, n
           if (t(ifail,ifail) == 0.0EXPON+0)
                return
       ifail = 0
       do j = 1, n {
           b(j) = b(j) / t(j,j)
           temp = -b(j)
           ADD(temp,[t(-,j)],b(-),j+1,n)
           }
       return
       end
```

With UNIX it is easy to set things up so that a command like

```
            version real inline trisol
```

produces version 1:

```
       SUBROUTINE TRISOL(T, LDT, N, B, IFAIL)
       INTEGER LDT, N, IFAIL
       REAL T(LDT, *), B(*)
*
*      TRISOL SOLVES LOWER TRIANGULAR LINEAR SYSTEMS OF
*      EQUATIONS.
*
*      ON ENTRY
*
*              T       REAL ARRAY
*                      T CONTAINS THE MATRIX OF THE SYSTEM.
*                      THE ZERO ELEMENTS OF THE MATRIX ARE
*                      NOT REFERENCED, AND THE CORRESPONDING
*                      ELEMENTS OF THE ARRAY CAN BE USED
*                      TO STORE OTHER INFORMATION.
*
*              LDT     INTEGER
*                      LDT IS THE LEADING DIMENSION OF THE
```

```
*                     ARRAY T.
*
*           N         INTEGER
*                     N IS THE ORDER OF THE SYSTEM.
*
*           B         REAL ARRAY
*                     B CONTAINS THE RIGHT-HAND SIDE OF THE
*                     SYSTEM.
*
*     ON RETURN
*
*           B         CONTAINS THE SOLUTION, IF IFAIL .EQ. 0.
*                     OTHERWISE B IS UNALTERED.
*
*           IFAIL     INTEGER
*                     IFAIL CONTAINS ZERO IF THE SYSTEM IS
*                     NONSINGULAR. OTHERWISE IFAIL CONTAINS
*                     THE SUBSCRIPT OF THE FIRST ZERO
*                     DIAGONAL ELEMENT OF T.
*
*     INTERNAL VARIABLES.
      INTEGER I, J
      REAL TEMP
      DO 1 IFAIL = 1, N
          IF (T(IFAIL, IFAIL) .EQ. 0.0E+0) RETURN
    1     CONTINUE
      IFAIL = 0
      DO 3 J = 1, N
          B(J) = B(J) / T(J, J)
          TEMP = -B(J)
          DO 2 I = J + 1, N
              B(I) = B(I) + TEMP*T(I, J)
    2         CONTINUE
    3     CONTINUE
      RETURN
      END
```

Similarly, the command

```
version dble calls trisol
```

might produce the desired version 4:

```
SUBROUTINE TRISOL(T, LDT, N, B, IFAIL)
INTEGER LDT, N, IFAIL
DOUBLE PRECISION T(LDT, *), B(*)
```

```
*
*      TRISOL SOLVES LOWER TRIANGULAR LINEAR SYSTEMS OF
*      EQUATIONS.
*
*      ON ENTRY
*
*            T      DOUBLE PRECISION ARRAY
*                   T CONTAINS THE MATRIX OF THE SYSTEM.
*                   THE ZERO ELEMENTS OF THE MATRIX ARE
*                   NOT REFERENCED, AND THE CORRESPONDING
*                   ELEMENTS OF THE ARRAY CAN BE USED
*                   TO STORE OTHER INFORMATION.
*
*            LDT    INTEGER
*                   LDT IS THE LEADING DIMENSION OF THE
*                   ARRAY T.
*
*            N      INTEGER
*                   N IS THE ORDER OF THE SYSTEM.
*
*            B      DOUBLE PRECISION ARRAY
*                   B CONTAINS THE RIGHT-HAND SIDE OF THE
*                   SYSTEM.
*
*      ON RETURN
*
*            B      CONTAINS THE SOLUTION, IF IFAIL .EQ. 0.
*                   OTHERWISE B IS UNALTERED.
*
*            IFAIL  INTEGER
*                   IFAIL CONTAINS ZERO IF IS
*                   NONSINGULAR. OTHERWISE IFAIL CONTAINS
*                   THE SUBSCRIPT OF THE FIRST ZERO
*                   DIAGONAL ELEMENT OF T.
*
*      INTERNAL VARIABLES.
       INTEGER J
       DOUBLE PRECISION TEMP
       DO 1 IFAIL = 1, N
           IF (T(IFAIL, IFAIL) .EQ. 0.0D+0) RETURN
    1      CONTINUE
       IFAIL = 0
       DO 2 J = 1, N
```

```
            B(J) = B(J) / T(J, J)
            TEMP = -B(J)
            CALL DADD(TEMP, T(1, J), B(1), J+1, N)
  2         CONTINUE
        RETURN
        END
```

In addition to the master file for *TRISOL*, one needs a few files of "macro definitions," that is, rules for generating program variants. The whole point of this approach is that these macro definitions can be used for generating variants of *any* procedure. For instance, suppose that we had 30 matrix procedures, each with the above four variants. Then a little over 30 files would suffice to maintain the 120 programs. In addition, this approach makes it substantially easier to keep all variants up to date when a program is changed for some reason.

(One reader conjectured that this lengthy example was being used for subconscious indoctrination to a particular style of commenting programs. That speculation was not entirely unjustified.)

PROGRAMMING ASSIGNMENTS

1. Implement Algorithm 4.2 as procedures *REDUCE* and *SOLVE*. [*Hint*: *REDUCE* and *SOLVE* will need a work array to communicate the row indices denoted by m in Algorithm 4.2.]

2. (Optional) Use whatever software tools you have available to mimic the use of the *M4* macro processor and the *EFL* translator that was described above.

EXERCISES

1. FORTRAN stores arrays by columns, i.e., in the order a_{11}, a_{21}, $a_{31}, \ldots, a_{12}, a_{22}, \ldots$. Given the declaration *REAL A(LDA,N)*, show that $A(I,J)$ is stored in location
$$LDA \times (J - 1) + I$$
of the region allotted to A.

2. Show that even when scaling is used to prevent overflow, computation of $(\sum v_i^2)^{\frac{1}{2}}$ produces a value that is correct to within a few ulps.

4.3 TESTING A LINEAR EQUATION SOLVER

How reliable is testing? In other words, if there is a mistake in your program, will testing find it? This section addresses these questions in the context of linear equation solvers.

We begin by presenting evidence for the claim that certain mistakes are easy to detect:

Hypothesis 4.1a
If a program that is intended to solve linear equations sometimes produces incorrect answers *in exact arithmetic*, then the error is likely to be exposed by almost any set of data.

While blunders can be caught by a naive approach to debugging, testing a linear equation solver for numerical oversights requires agreement on a testable specification. We will use the following:

Specification 4.1 (Informal)
The computed solution of *A*x = b should be such that *A*x is very close to b, relative to the computer's precision.

A number of linear equation solvers are known to satisfy the formal statement of this specification that is given later in this section. However, the algorithms in Section 4.1 fail to satisfy it, though they differ in important ways in the nature and extent of their failure.

The third goal of this section is to present evidence for the claim that numerical oversights are difficult to detect:

Hypothesis 4.1b
Many numerically unstable linear equation solvers can be exposed by testing, but it is generally inadequate to use a small number of randomly chosen sets of test data.

Finally, we discuss some systematic approaches to generating test data.

4.3.1 Detecting Blunders

We will argue for Hypothesis 4.1a on both theoretical and experimental grounds.

The Theoretical Justification. Let us begin by illustrating our reasoning in the setting of polynomial evaluation, which is a much simpler context than that of linear equations.

First, an example. Suppose you mistakenly type the one-line program

$$p \leftarrow (x + 1) + x$$

instead of

$$p \leftarrow (x + 1) \times x$$

and then test the program with a single input x. The mistake will be revealed unless $2x + 1 = x^2 + x$, that is, unless x equals either $x_1 = (1 - \sqrt{5})/2$ or $x_2 = (1 + \sqrt{5})/2$. See Figure 4.3.

Sec. 4.3   Testing a Linear Equation Solver   95

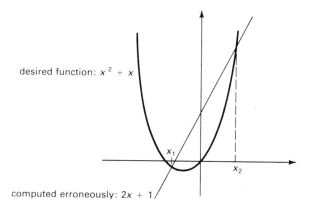

Figure 4.3

In general, suppose we are trying to write a program that evaluates a polynomial $p(x)$ of degree n [that is, $p(x) = a_n x^n + a_{n-1} x^{n-1} + \cdots$]. As long as our program applies the operations $+$, $-$, and $\times$ to the datum x and various constants, the program computes *some* polynomial $q(x)$. Suppose that the program is incorrect, so that $q(x) \neq p(x)$ for at least one input x. A well-known theorem of mathematics implies that there are at most k inputs x such that $q(x)$ equals $p(x)$, where k is the maximum degree of the two polynomials. See Figure 4.4. Thus running the program with any input other than one of a handful $x_1, x_2, \ldots, x_k$ of unfortunate sets of test data will produce an incorrect result and thereby expose the error.

This reasoning extends to the problem of solving linear equations. As shown by Algorithm 4.1a, linear equations can be solved using only the arithmetic operations of addition, subtraction, multiplication, and division. Thus the solution **x** is a *rational* function of the data $A|\mathbf{b}$. There exist mathematical theorems showing that if a would-be linear equation solver computes a *different* rational function, then the computed function differs from the rational function defined by the linear equation problem on "almost every" set of data. In fact, the argument can be extended to include programs that use test-and-branch statements. [See "An Analysis of the Effect of Rounding Error on the Flow of Control of Numerical Processes" by Esko Ukkonen (*BIT*, 1979, pp. 116–133).] In other words, an algorithm that consists of applying the arithmetic operations $+$, $-$, $\times$, and $/$, with execution

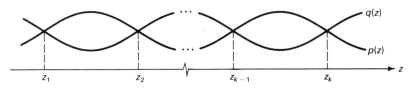

Figure 4.4

controlled by comparisons between real values, either (1) solves linear equations or (2) fails to solve them for almost every set of data. In the latter case, almost any set of data will expose the error.

Because the above argument is predicated on the use of exact arithmetic, its applicability in practice is necessarily limited. For instance, the discussion about polynomial evaluation rests on a theorem implying that a change to the program necessarily changes all but a few of its computed results. Where fixed-precision arithmetic is involved, this theorem can be misleading. For example, rather extensive changes to the coefficients of the approximating polynomial $p(x)$ leave the computed results of the sine function unaffected. (See Exercise 1 of Section 3.3 and Exercise 1 below.)

A Mutation Experiment. The following experiment was motivated by the limited applicability of the above theoretical justification and provides empirical evidence for Hypothesis 4.1a. The intuitive thrust of the experiment is that a typographical mistake in a linear equation solver will probably be caught by even a naive debugging strategy, unless the mistake's only effect on the program's behavior is to make it numerically unstable or inefficient.

The programs used in our experiment were straightforward FORTRAN implementations of

Algorithm 4.1a (Gaussian elimination without pivoting),

Algorithm 4.1b (Gaussian elimination with partial pivoting), and

Algorithm 4.1c (Gauss-Jordan elimination with partial pivoting).

The programs contained absolutely no embellishments, not even to guard against division by zero.

These three programs were subjected to a mutation experiment, as described in Section 1.1. We ran the mutants on a single set of data produced by first generating A and the "true solution" **x** using

for $i = 1$ to n

for $j = 1$ to n

$a_{ij} \leftarrow random(-1,1)$

$x_i \leftarrow random(-1,1)$

and then computing **b** as $A \times \mathbf{x}$. [Here $random(-1, 1)$ denotes a "random" value between -1 and 1; specifically, we used $2 \times ran(\ ) - 1$, where $ran(\ )$ is the random number generator of Section 1.1.] Our criterion for accepting a computed solution was that each of its entries should lie within 10^{-4} of the true value.

The only mutants that survived were those that correctly solve linear equations *in exact arithmetic*. Specifically, the only typographical mistakes that went undetected were those that cause Algorithms 4.1b and 4.1c to select improper pivot rows (resulting in a numerically unstable program) and a handful of other mistakes (see Exercise 2) that do not affect the computed solution (even when fixed-precision arithmetic is used).

4.3.2 A More Formal Specification

As Wilkinson observed in the 1940s (see Section 2.4), the better linear equation solvers produce a solution x that "almost" solves the given equations. However, turning this observation into a formal specification is difficult. There are a number of ways to formalize Specification 4.1, but most approaches will not work because they result in a specification that is not satisfied, even by the best algorithms. Discovery of an appropriate specification involves theoretical analysis of error propagation, just as with the sine function (Chapter 3) and root refining procedures (Chapter 5).

For a concise statement of the specification we will employ the following notation. If Z is a matrix or a vector, we will use $|Z|$ to denote the largest absolute value of Z's entries. Thus if

$$A = \begin{bmatrix} -8 & 7 \\ -5 & 6 \end{bmatrix},$$

then $|A| = 8$.

Data $A|\mathbf{b}$ will be assigned the "instability score"

$$\omega(A|\mathbf{b}) = \frac{|A\mathbf{x} - \mathbf{b}|}{|A| \times |\mathbf{x}| \times \epsilon},$$

where $\mathbf{x}$ is the computed solution of $A\mathbf{x} = \mathbf{b}$ and ϵ denotes machine epsilon. Using an example from Section 4.1, suppose that Gaussian elimination (with or without pivoting) is used to solve the linear equations $A\mathbf{x} = \mathbf{b}$, where

$$A = \begin{bmatrix} 0.578 & 0.323 \\ 0.377 & 0.212 \end{bmatrix}, \quad \mathbf{b} = \begin{bmatrix} 0.901 \\ 0.589 \end{bmatrix}$$

The solution computed in three-decimal arithmetic is

$$\mathbf{x} = \begin{bmatrix} 0.441 \\ 2.00 \end{bmatrix},$$

so

$$A\mathbf{x} - \mathbf{b} = \begin{bmatrix} 0.900898 \\ 0.590257 \end{bmatrix} - \begin{bmatrix} 0.901 \\ 0.589 \end{bmatrix} = \begin{bmatrix} -0.000102 \\ 0.001257 \end{bmatrix}$$

Thus

$$\omega(A|\mathbf{b}) = \frac{0.001257}{0.578 \times 2 \times 0.01} = 0.108\ldots$$

We are ready to formalize our specification:

Specification 4.1 (Formal)
Consider a given linear equation solver and given data $A|\mathbf{b}$ for which the procedure computes a solution $\mathbf{x}$. Define

$$\omega(A|\mathbf{b}) = \frac{|\mathbf{r}|}{|A| \times |\mathbf{x}| \times \epsilon}$$

where $\mathbf{r}$ is the residual vector

$$\mathbf{r} = A\mathbf{x} - \mathbf{b}$$

and ϵ denotes machine epsilon.

   a. For each value of n there should exist a number $bound(n)$ with the property that $\omega(A|\mathbf{b}) \leq bound(n)$ for all $n \times n$ sets of data for which a solution is computed without underflow.
   b. In addition, $bound(n)$ should be a slowly growing function of n.

Specification 4.1 is intentionally imprecise about how large ω is allowed to be. It makes little sense to say, for example, that if $n = 4$, then $\omega = 7.1$ is OK, while $\omega = 7.2$ is unacceptable. In practice, experience indicates that one can interpret part a as the requirement that $\omega \leq 20$ for all 4×4 linear equations. Furthermore, most linear equation solvers that score $\omega > 10$ on one data set will score $\omega > 100$ on another data set, so the *20* can be changed to, say, *10* or *100* without affecting which programs satisfy the condition.

In the paper "Numerical Linear Algebra on Digital Computers" (*Bulletin of the Institute of Mathematics and its Applications*, 1974, p. 356) James Wilkinson makes the following remarks. Although these comments were originally made in the context of rigorous proofs that a matrix algorithm meets certain specifications, they are equally appropriate for testing.

> On the debit side there does seem to be some misunderstanding about the purpose of *a priori* backward error analysis. All too often, too much attention is paid to the precise error bound that has been established. The main purpose of such an analysis is either to establish the essential numerical stability of an algorithm or to show why it is unstable and in doing so to expose what sort of change is necessary to make it stable. The precise error bound is not of great importance.

With regard to condition b, it should be adequate in practice to require either $\omega \leq 10^3$ or $\omega \leq 10^4$ when $n = 30$. The main subtleties with Specification 4.1 involve linear equation solvers that satisfy part a but not part b. As it turns out, Algorithms 4.1a and 4.1c fail to satisfy part a of Specification 4.1, while Algorithm 4.1b satisfies part a but not part b.

While it is difficult to see exactly why ω was defined this way, it is not hard to see that the definition is at least plausible. The following discussion

Sec. 4.3 Testing a Linear Equation Solver

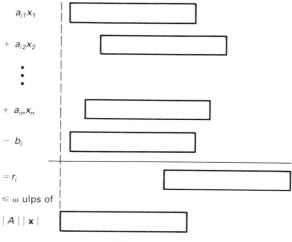

Figure 4.5

indicates that a small value of ω guarantees that almost all figures cancel when the residual $\mathbf{r}$ is computed. Writing the definition of $\omega = \omega(A|\mathbf{b})$ as $|\mathbf{r}| = \omega |A| |\mathbf{x}| \epsilon$ and remembering that $|A| |\mathbf{x}| \epsilon$ is approximately the spacing of floating-point numbers around $|A| |\mathbf{x}|$ (recall the most important fact about floating-point number systems), we see that the largest entry of $\mathbf{r}$ is approximately the size of ω ulps of $|A| |\mathbf{x}|$. Furthermore, since $|a_{ij} x_j| \leq |A| |\mathbf{x}|$ for all i and j, it follows that computation of the ith entry of $\mathbf{r}$ must have the appearance shown in Figure 4.5.

Discussion. We are lucky to have a plausible and testable specification for linear equation solvers. There exist computational problems for which algorithms can be proved to satisfy a certain specification in theory but for which methods of verification in practice are more complex and dubious than the software being tested. In particular, this is true of the "linear least squares problem," i.e., the problem of solving overdetermined linear systems $A\mathbf{x} = \mathbf{b}$ having more equations than unknowns (the "solution" minimizes the sum of squares of entries in the residual vector $\mathbf{r}$). For a discussion of this point, see pp. 5 and 6 of the paper "Research, Development and LINPACK" by G. W. Stewart (in *Mathematical Software III*, edited by John R. Rice, Academic Press, New York, 1977). At the end of this section we will discuss an approach that avoids this difficulty by measuring rounding errors' effect on an idealized computation, rather than on a computation by a particular machine.

4.3.3 The Difficulty of Testing Linear Equation Solvers

As discussed here, testing a linear equation solver involves generating a sequence $A_1|\mathbf{b}_1, A_2|\mathbf{b}_2, \ldots, A_i|\mathbf{b}_i, \ldots,$ of test data sets, computing the corresponding solutions $\mathbf{x}_i$ using the program being tested, and evaluating the acceptability of each $\mathbf{x}_i$ using $\omega(A_i|\mathbf{b}_i)$. It is simple and often effective to use data generated as follows. [As before, *random*$(-1,1)$ denotes a "random" value between -1 and 1.]

Data 4.1a
..Generate linear equations.

for $i = 1$ to n

    for $j = 1$ to n

        $a_{ij} \leftarrow$ *random*$(-1,1)$

    $b_i \leftarrow$ *random*$(-1,1)$

General theorems discussed in the book *Software for Roundoff Analysis of Matrix Algorithms* by Webb Miller and Celia Wrathall (Academic Press, New York, 1980) imply that tests of linear equation solvers may as well be restricted to sets of data with entries between -1 and 1; numerical instabilities, if they exist, will be exhibited by such data. This happens because the general effect of rounding errors is preserved if all entries of the data are multiplied by a constant. (Specifically, see the observations about "homogeneous" algorithms on pp. 70, 82, 90, and 94 of that book.)

An Experiment. To investigate Hypothesis 4.1b, we performed the following experiment to assess the effectiveness of testing as a means of deciding whether a linear equation solver satisfies part a of Specification 4.1:

Experiment 4.1a
repeat 5000 times

    $A|\mathbf{b} \leftarrow$ Data 4.1a with $n = 4$.

    for each of several numerically unstable linear equation solvers

        $\mathbf{x} \leftarrow$ apply the linear equation solver to $A|\mathbf{b}$.

        Make a record of $\omega(A|\mathbf{b})$.

Nine linear equation solvers were used in the experiment. All nine fail to satisfy part a of Specification 4.1. The reader is welcome to skip over the following list of references for the programs used in the experiment.

In addition to implementations of Algorithms 4.1a and 4.1c, we used implementations of the following procedures. All of them are, in fact, specialized versions of linear least squares procedures.

Sec. 4.3   Testing a Linear Equation Solver   101

C The classical Gram-Schmidt method. See p. 204 of *Numerical Methods* by Germund Dahlquist, Ake Björck, and Ned Anderson (Prentice-Hall, Englewood Cliffs, N.J., 1974).

F An unstable variant of the Fast Givens method. See the formulas labeled (9′) on p. 332 of "Least Squares Computations by Givens Transformations Without Square Roots" by W. Morven Gentleman (*Journal of the Institute of Mathematics and its Applications*, 1973).

G Unstable plane rotations. See p. 239 of *The Algebraic Eigenvalue Problem* by James Wilkinson (Oxford University Press, London, 1965). Using $e = \sin\vartheta/(1 + \cos\vartheta)$, the inner loop can be accomplished in only three multiplications by $t \leftarrow a_{rj}$; $a_{rj} \leftarrow t \times \cos\vartheta + a_{ij} \times \sin\vartheta$; $a_{ij} \leftarrow e \times (t + a_{rj}) - a_{ij}$.

H Householder's method without choice of signs. See pp. 233–236 of Wilkinson's book. In formula (38.3) on p. 153, k is always chosen as $-S$.

M Application of the modified Gram-Schmidt method to A, with application of the classical Gram-Schmidt method to **b**. See pp. 201–204 of the book by Dahlquist, Björck, and Anderson.

N The method of normal equations. See pp. 220–221 of the book by Dahlquist, Björck, and Anderson.

S Unstable storage of plane rotations. See p. 239 of Wilkinson's book, especially the last two sentences of section 50. **b** is transformed using $\cos\theta$ as reconstructed by the formula $\pm\sqrt{1 - \sin^2\theta}$.

The values of ω that we recorded for Experiment 4.1a are summarized in the following table:

	Algorithm							
	4.1a	4.1c	C	F	G	H	M	N	S
$\omega \leq 5$	4402	4996	4881	4765	4607	4890	4931	4036	4839
$5 < \omega \leq 20$	454	4	110	180	304	88	64	713	126
$20 < \omega \leq 100$	124	0	9	50	71	18	5	195	33
$100 < \omega$	20	0	0	5	18	4	0	56	2

Except for Algorithm 4.1c, the instability signal $\omega > 20$ appeared for every algorithm. However, note that a randomly chosen handful of test cases may not reveal the failures. For instance, Algorithm 4.1a scored over 5 less than 12% of the time, so running it on, say, 10 sets of data may well not turn up anything suspicious.

Notice that even extensive testing may miss the fact that Algorithm 4.1c does not satisfy part a of Specification 4.1. The largest of the 5000 values of ω discovered for that algorithm was about 7.1. While it is impossible to draw a sharp line between acceptable and unacceptable value of ω, the value 7.1 is probably not so large that it would raise many eyebrows. The failure of Algorithm 4.1c can be observed, however, with the data $A|\mathbf{b}$ =

0.58173	−0.94685	0.45988	−0.86058	0.49692
−0.05527	0.09006	0.42733	−0.25186	0.76711
−0.02706	0.04404	0.25181	0.13112	0.49512
0.01371	−0.02231	0.10565	−0.76055	0.06692

For this set of data, which was located by a procedure discussed at the end of this section, Algorithm 4.1c scores $\omega > 6000$ on each of the three machines we tried. Perhaps the difficulty of detecting the instability of Algorithm 4.1c is related to that fact that the algorithm does satisfy a specification that is just slightly weaker than Specification 4.1. [See "On the Stability of Gauss-Jordan Elimination with Partial Pivoting" by G. Peters and J. H. Wilkinson (*Communications of ACM*, Jan. 1975, pp. 20–24).]

At several points in this book we have asserted that it may require a considerable amount of testing to detect numerical instability in a program. (For instance, see Example 1 of Section 2.4.) Thus the conclusion of Experiment 4.1a should come as no surprise. However, there exist examples of numerical instabilities, in algorithms for other computational problems, that are exposed by almost any set of test data. See Section 5.4.

4.3.4 Systematic Generation of Test Data

In some cases, ill-conditioned test data (see Section 4.1) are more effective at exposing a numerical instability than are purely random linear equations (Data 4.1a). For instance, with algorithms C and N (see above), average values of ω are much greater for ill-conditioned data; with algorithm M they are slightly greater. However, experience indicates that ill-conditioned linear equations are, on the whole, relatively inefficient for diagnosing instabilities. See Programming Assignment 3.

Another systematic approach to testing is to keep track of the worst test data (i.e., where ω is largest) and generate linear equations that lie near this worst data set. For instance, a random number generator might be used to make small random changes to the worst known equations.

A sophisticated variation of this approach follows from viewing the process of testing linear equation solvers as a numerical optimization problem. For $n = 4$, the problem is as follows: *Can the 20 entries of $A|\mathbf{b}$ be adjusted so that the "objective function" $\omega(A|\mathbf{b})$ is larger than, say, 100?* Numerical optimization programs, sometimes called "hill-climbing" routines, can be used to automatically alter the entries of $A|\mathbf{b}$ in a systematic search for large values of ω. A number of numerical optimization programs are available, but none works efficiently unless the objective function is well behaved. This observation makes the following hypothesis plausible:

Hypothesis 4.1c
The erratic behavior of ω complicates systematic generation of effective test data for linear equation solvers.

A procedure for computing a "smooth" approximation to ω is described in the book *Software for Roundoff Analysis of Matrix Algorithms* by Webb Miller and Celia Wrathall (Academic Press, New York, 1980). The procedure is too complex to be repeated here, but the relationship between ω and this smooth approximation $\bar{\omega}$ can be pictured intuitively as shown in Figure 4.6.

The results of the following experiment indicate that the effectiveness of systematic search techniques for testing linear equation solvers is strongly influenced by the smoothness of the function used to score the accuracy of the computed solution.

Experiment 4.1b
repeat 10 times

    $A|\mathbf{b} \leftarrow$ Data 4.1a with $n = 4$.

    for each of several numerically unstable linear equation solvers

        Apply a numerical "hill-climbing" routine to ω. Begin at $A|\mathbf{b}$ and record *success* if a value $\omega > 100$ is found.

        Apply the same "hill-climbing" routine to $\bar{\omega}$. Begin at $A|\mathbf{b}$ and record *success* if a value $\omega > 100$ is found.

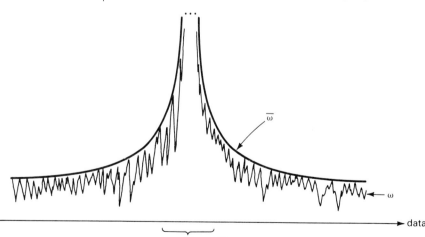

Figure 4.6

As the following table shows, the maximizer was substantially more effective on $\bar{\omega}$:

	Algorithm								
	4.1a	4.1c	C	F	G	H	M	N	S
Successes with ω	3	0	0	2	3	1	0	2	2
Successes with $\bar{\omega}$	10	4	10	10	8	8	10	10	10

The hill-climbing procedure was constructed to stop after 500 samples of the objective function. Thus, in the course of being applied to $\bar{\omega}$ with 10 randomly generated starting points, it evaluated $\bar{\omega}$ at most 5000 times, yet was able to find values of ω for Algorithm 4.1c in excess of 100. When applied directly to ω, the maximizer had no such luck: The largest value found was $\omega \approx 2.6$, less than the $\omega \approx 7.1$ found using 5000 random sets of data (Experiment 4.1a).

Using $\bar{\omega}$ in the "hill-climbing" approach to generating test data increases computer costs at least a factor of 10. However, this penalty is offset by the following advantages:

1. The process of evaluating $\bar{\omega}$ proceeds automatically from the statement of a matrix algorithm; it is not necessary to supply a reliable procedure to score (analogous to ω) the computed solution. Thus the process works for general linear least squares procedures just as well as for linear equation solvers.
2. The extra work done in evaluating $\bar{\omega}$ produces information that often either points the way to a numerically stable program or shows that no simple modification of the algorithm is stable.

The interested reader is referred to the book by Miller and Wrathall for a detailed development of this approach to testing matrix algorithms.

PROGRAMMING ASSIGNMENTS

1. Test whether your linear equation solver satisfies part a of Specification 4.1 when $n = 4$.

    Warnings: The data $A|\mathbf{b}$ must be saved to compute ω. The subtraction $A\mathbf{x} - \mathbf{b}$ suffers extreme cancellation and should be performed using extra precision, or the result may be misleading. If A, B, X, RI, and $RMAX$ are single precision and if SUM is double precision, then $RMAX = |\mathbf{r}|$ should be computed as follows:

    ```
    N = 4
    RMAX = 0.0
    DO 100 I = 1, N
        SUM = 0.0D0
        DO 90 J = 1, N
    ```

```
            SUM = SUM + DBLE(A(I,J))*DBLE(X(J))
 90      CONTINUE
         RI = SUM - DBLE(B(I))
         RMAX = MAX(RMAX,ABS(RI))
100   CONTINUE
```

If A, B, and X are already in the highest hardware precision, then one might use software techniques to simulate even higher precision. To see how this can be done, consult the paper "A Fortran Multiple-Precision Arithmetic Package" by Richard Brent (*ACM Transactions on Mathematical Software*, March 1978, pp. 57–70) or "Software for Doubled-Precision Floating-Point Computations" by Seppo Linnainmaa (*ACM Transactions on Mathematical Software*, Sept. 1981, pp. 272–283.)

The lack of higher precision is not necessarily disastrous. Our experience suggests that computation of $A\mathbf{x} - \mathbf{b}$ in working precision will probably not contaminate the computed value of ω enough to make any difference. One reason is that all we really need are the first few bits of ω. Moreover, the values of ω that interest us are the large ones, where, by definition, at least a few digits do not cancel in the computation of $\mathbf{r}$.

2. (Optional) Try other ways of formalizing Specification 4.1 and investigate whether they hold in practice.

3. (Optional) Test your linear equation solver with data that are guaranteed to be ill-conditioned. In particular, use Data 4.1b in place of Data 4.1a, perhaps picking $n = 4$ and $K = 2$.

The first K columns of A, call them $\mathbf{a}_1, \ldots, \mathbf{a}_K$, are generated randomly. Each of the remaining columns is found as follows. The columns $\mathbf{a}_1, \ldots, \mathbf{a}_K$ are scaled by constants, the resulting vectors are added, and a little "noise" is thrown in for good measure.

Data 4.1b

..Generate ill-conditioned linear equations.

tiny ← 0.001   ... pick another number if you like

K ← some integer less than n

..First generate K columns of A and generate **b**.

for $i = 1$ to n

    for $j = 1$ to K

        a_{ij} ← *random*(− 1,1)

    b_i ← *random*(− 1,1)

..Generate each of the remaining columns of A so as to
..lie near a linear combination of columns 1 to K.

    for $j = K + 1$ to n

    ..Get coefficients for the linear combination.

    for $k = 1$ to K

        $c_k \leftarrow$ random$(-1,1)$

    for $i = 1$ to n

        $a_{ij} \leftarrow \sum_{k=1}^{K} c_k \times a_{ik} +$ tiny $\times$ random$(-1,1)$

EXERCISES

1. Suppose that the range of floating-point exponents is approximately symmetric around 0, i.e., $m \approx -M$ in the notation of Section 2.1. Show that the computed values of the polynomials $p(x) = x^2$ and $q(x) = 0$ (the polynomial whose value is always 0) are identical for roughly 25% of all floating-point numbers x. Relate this observation to the theoretical justification for Hypothesis 4.1a.

2. Consider the innermost *for* statement

$$for\ j = k + 1\ to\ n$$

of Algorithms 4.1a, 4.1b, and 4.1c. Show that changing $k + 1$ to k, 1, $k + 0$, $k/1$, $k * 1$, or $k**1$ constitutes a *mistake* but not an *error* (in the sense of Section 1.1). How do these mistakes make their presence known when the programs are executed? (This trickle of examples becomes a flood in the next chapter. See the discussion of a mutation experiment in Section 5.2.)

3. Let x' be the value computed by

$$\pi \leftarrow 3.14$$
$$x \leftarrow t - 4 \times \pi$$

in three-decimal arithmetic, and define $\omega(t) =$ ulps(x', x), where x denotes the true value. Thus if $t = 13.0$, then $x' = 0.4$, $x = 0.44$, and $\omega(t) = 40$.
   (a) Show that $\omega(12.6) = \infty$.
   (b) Show that if t is a representable number satisfying $13.6 \leq t \leq 22.5$, then $\omega(t) = 4$. Thus samples of $\omega(t)$ for t between 13.6 and 22.5 give no indication of how to vary t to look for large values of $\omega(t)$.
   (c) An argument in Section 3.3 justifies the approximation $\omega(t) \approx \bar{\omega}(t) = |t/x|$ when t has the same order of magnitude as 4π. Sketch $\bar{\omega}(t)$ and argue informally that two samples of $\bar{\omega}(t)$, say with $13.6 \leq t \leq 22.5$, will indicate that t should be decreased to look for large values of $\omega(t)$.

4. (Optional) How would a mistake in the computation of Data 4.1b affect tests conducted for Programming Assignment 3? How would you detect it?

RESEARCH PROJECT

Find a highly reliable testing procedure for linear equation solvers.

4.4 A PROPERTY THAT ALMOST ALWAYS HOLDS

In some circumstances it is possible to realize a significant savings in computer time by using a program whose output almost always possesses a certain property, instead of a program whose results can be guaranteed. In particular, that phenomenon is exhibited by Gaussian elimination with partial pivoting, which almost always satisfies Specification 4.1; specifically, solutions of $n \times n$ equations computed by that method almost always score $\omega < n$. In this section we will explain in more detail the manner in which Gaussian elimination with partial pivoting violates Specification 4.1. Then we will discuss the evidence supporting the claim that the failure does not occur in practice and explore the cost of using a linear equation solver that is guaranteed to meet our specification.

Failure in Theory. Theoretical analyses of the effect of rounding errors on Gaussian elimination show that the size of the instability measure ω is determined by the growth in the sizes of the entries of A during the triangularization phase. Given this observation, it is possible to concoct $n \times n$ linear equations where Gaussian elimination with partial pivoting produces an instability measure $\omega(A|\mathbf{b})$ approaching 2^n. For instance, taking A to be the 30×30 version of the matrix

$$\begin{bmatrix} 1 & 0 & 0 & 0 & 1 \\ -1 & 1 & 0 & 0 & 1 \\ -1 & -1 & 1 & 0 & 1 \\ -1 & -1 & -1 & 1 & 1 \\ -1 & -1 & -1 & -1 & 1 \end{bmatrix}$$

(that is, $a_{ij} = -1$, 1, or 0 according to whether $j < i$, j equals either i or n, or $i < j < n$) and $b_i \leftarrow ran()$ for $1 \leq i \leq 30$, we found $\omega(A|\mathbf{b}) \approx 10^7$, in spite of the fact that no rounding errors are committed in reducing A to triangular form. Thus Gaussian elimination with partial pivoting does not satisfy part b of Specification 4.1, since 2^n cannot be considered a "slowly growing function of n."

Success in Practice. The failure occurs only under extremely rare conditions. No proof of this fact has been given; it is not even clear what it might mean for Specification 4.1 to "almost always" hold. Still, empirical evidence has convinced the experts to trust partial pivoting. The evidence consists of (1) observations made on sets of data that arise in practice and (2) measurements made using artificially generated sets of data.

For example, we measured Gaussian elimination with partial pivoting by applying the procedure followed in Experiment 4.1a, this time with $n = 30$ instead of $n = 4$. (You should not try this unless you have a generous

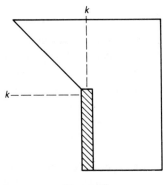
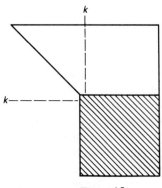

Figure 4.7 **Figure 4.8**

computer budget.) The largest of the 5000 values of ω was about 21, which supports the claim that $\omega(A|\mathbf{b})$ is almost always less than n.

The Cost of Perfection. Other linear equation solvers can be *guaranteed* to satisfy Specification 4.1 [typically where *bound*(n) is taken to be a low-degree polynomial in n]. However, unlike the addition of partial pivoting to Gaussian elimination, the penalty in computer time is appreciable, perhaps an additional 50% or more. Gaussian elimination with *complete* pivoting is one such guaranteed procedure. Instead of picking the kth divisor as the largest entry in the portion of A indicated in Figure 4.7 and interchanging rows to bring the divisor up to the k, k position, the kth divisor is chosen to be the largest entry in the region shown in Figure 4.8 and rows *and columns* are interchanged. Complete pivoting adds significantly to execution costs because the number of required comparisons is about equal to the number of floating-point multiplications used by Gaussian elimination. (This follows from the observation that the process of introducing zeros into column k costs roughly one comparison and one multiplication for each entry in the shaded square depicted in Figure 4.8.)

EXERCISE

1. Let Gaussian elimination with partial pivoting be applied to the $n \times n$ matrix of the form

$$\begin{bmatrix} 1 & 0 & 0 & 0 & 1 \\ -1 & 1 & 0 & 0 & 1 \\ -1 & -1 & 1 & 0 & 1 \\ -1 & -1 & -1 & 1 & 1 \\ -1 & -1 & -1 & -1 & 1 \end{bmatrix}$$

Show that no rounding errors are committed (if we neglect **b**) and that the n, n-entry of the reduced matrix is 2^{n-1}.

5
SOLVING A NONLINEAR EQUATION

The main goals of this chapter are (1) to discuss some general characteristics of performance measurement and (2) to explore the boundary between numerical problems that can be completely solved by a computational procedure and those that cannot. To do this, we will consider the solution of a single nonlinear equation, a simple problem that in some sense lies near the boundary.

The mathematical formulation of the problem of solving a nonlinear equation is as follows: Given a function f, determine a number x such that $f(x) = 0$. Such a number x is called a *zero* or *root* of f. For numerical computation this formulation is inadequate because even if f is "smooth and crosses the x-axis," there may be no floating-point number x for which the computed value of $f(x)$ is exactly zero. Intuitively speaking, the zero of f may fall between adjacent representable numbers.

Instead, we will deal with the problem of locating either a zero of f or a tiny interval on which f changes sign. Let us define a *sign-change* of f to be a pair a, b of representable numbers such that either (1) $a = b$ and $f(a) = 0$ or (2) $a < b$ and $sign(f(a)) \neq sign(f(b))$. We will think of $sign(0)$ as undefined, so implicit in (2) is the requirement that $f(a)$ and $f(b)$ are nonzero. The problem of solving a nonlinear equation has a two-part formulation that is appropriate for numerical computation:

The Root Bracketing Problem
Let f be a function determined by a particular computational procedure. Find a sign-change of f. If none exists, then report that the problem has no solution.

The Root Refining Problem
Let *f* be a function determined by a particular computational procedure and let *a*, *b* be a given sign-change of *f*. Find a sign-change *a'*, *b'* of *f* such that *a'* and *b'* lie between *a* and *b* and differ by at most one ulp.

As we saw in Section 1.3, the root bracketing problem cannot be solved by a foolproof computational procedure.

Section 5.1 presents a simple algorithm that essentially solves the root refining problem and discusses the testing of implementations of that algorithm. In Section 5.2 we develop a modified algorithm based on linear interpolation (sometimes called the Regula Falsi method or method of false position) that appears to be more efficient when the graph of f is smooth. Measuring the efficiency of such procedures is the topic of Section 5.3. We will see that careful measurements of performance can reveal implementation mistakes and expose inadequacies in the underlying algorithm that are not revealed by testing. In particular, measurements reveal that rounding errors diminish efficiency of the algorithm developed in Section 5.2. (Rounding errors have no adverse effect on its accuracy!) Means of correcting this inefficiency are presented.

In summary, the approach taken in Sections 5.1–5.3 is to develop an efficient and completely reliable procedure that solves a restricted formulation of a numerical problem that is, in its full generality, unsolvable. This approach has been less effective for other unsolvable problems. Part (though by no means all) of the reason for the failure of this approach in other areas hinges on the use of fixed-precision arithmetic. The main goal of Section 5.4 (which is optional) is to investigate this category of difficulties, i.e., to show that the use of floating-point arithmetic poses obstacles for attempts to solve special cases of unsolvable numerical problems. In particular, we will observe that inherent properties of numerical computation cause several difficulties when we try to repeat our success with nonlinear equations in the closely related context of minimizing a function.

5.1 BISECTION

The following algorithm essentially solves the root refining problem. It repeatedly bisects the interval $[a, b]$, at each step retaining a half on which f changes sign. The process terminates with a and b either set to a root of f or altered so that the computed midpoint $(a + b)/2$ does not lie strictly between them.

Algorithm 5.1

..If at any time a computed value $f(x)$ is zero,

..set $a \leftarrow x$, $b \leftarrow x$, and return.

repeat

$$c \leftarrow \frac{a+b}{2}$$

if $c \leq a$ or $c \geq b$

done

if $sign(f(a)) = sign(f(c))$

$a \leftarrow c$

else

$b \leftarrow c$

Notice that since Algorithm 5.1 terminates upon finding a root of f, its behavior does not depend on how we interpret $sign(0)$.

In many respects Algorithm 5.1 is about as accurate a root refining procedure as one might hope to find; it is guaranteed to refine a sign-change of f to a very small interval. The following specification makes this property precise.

Specification 5.1

Let f, a, and b be given, where a,b is a sign-change of f. Assume that the root-refining procedure is executed on a binary (i.e., $\beta = 2$) machine and terminates without overflow. The procedure should find a sign-change a', b' of f such that $a \leq a' \leq b' \leq b$ and either (1) $ulps(a',b') \leq 1$ or (2) one of a' or b' is zero and the other has magnitude less than 2σ, where σ is the smallest positive floating-point number.

As in Chapters 3 and 4, the specification was discovered by analyzing an algorithm. (See especially the introduction to Chapter 3.) In particular, discovery of the analysis of Algorithm 5.1 given below led us to refine the specification to include mention of the floating-point base and pay particular attention to the underflow threshold σ.

To verify that Algorithm 5.1 meets Specification 5.1, three things can be shown:

A. Algorithm 5.1 terminates. (We are not interested in procedures that satisfy Specification 5.1 "vacuously" by getting stuck in an infinite loop.)

B. The values of a and b move toward one another, maintaining the property of being a sign-change of f.

C. If $\beta = 2$ and overflow does not occur, then one of the conditions (1) or (2) holds.

Verification of properties A and B is left as Exercises 3 and 4. We will now verify property C.

If Algorithm 5.1 stumbles across a root of f, then it returns with $a = b$, so (1) holds. It follows that we need only consider the case that the final values of a, b, c satisfy either $c \leq a$ or $c \geq b$. We need to consider two cases.

First suppose that neither a nor b is zero. We will show that condition (1) holds; that is, a and b are at most an ulp apart. Our method of proof will be to assume that $ulps(a, b) > 1$ and then derive a contradiction. With that in mind, assume that a and b are more than one ulp apart but that the computed value of c satisfies either $c \leq a$ or $c \geq b$. In particular, assume that $c \geq b$, since the reasoning when $c \leq a$ is essentially the same. If a is more than two ulps from b, then change a to make it exactly two ulps less than b; this cannot affect the fact that the computed midpoint is at least b. Let y be the floating-point number between a and b; our contradiction will be that the computed value of $(a + b)/2$ is y. If $|y|$ is not a power of 2, then a and b are equidistant from y and $a + b = 2y$. It follows that $(a + b)/2 = y$ is computed exactly. On the other hand, suppose that $|y| = 2^k$. Without loss of generality we can assume that a and b are positive, so $a < 2^k < b$. If $\delta = 2^k - a$, then $b = 2^k + 2\delta$. (Remember that the spacings on either side of a power of β differ by the factor β.) See Figure 5.1. It follows that $a + b = (2^k - \delta) + (2^k + 2\delta) = 2^{k+1} + \delta$, so the computed value of $a + b$ is 2^{k+1}, and y is the computed midpoint. In either case, we find that the assumption $ulps(a, b) > 1$ leads to the conclusion that $a < c < b$, thereby contradicting the hypothesis that $c \geq b$.

Now suppose that one of a or b is zero, say $a = 0$. We will show that condition (2) holds. The computed value of $(a + b)/2$ is either $b/2$ or, because of underflow, zero. If the value of $b/2$ is representable, then the algorithm repeats. If the algorithm terminates at this point, then underflow has occurred, so $|b/2| < \sigma$, that is, $|b| < 2\sigma$. This completes the verification that Algorithm 5.1 satisfies Specification 5.1.

Testing. An implementation of Algorithm 5.1, or of one of the variants developed later, should be tested to see whether it satisfies Specification 5.1.

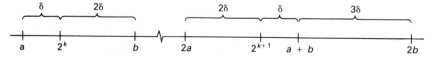

Figure 5.1

Sec. 5.1   Bisection   113

The reasoning given above "proves" that the specification will be met, but testing is needed because (1) our assumptions about floating-point arithmetic may be violated in a way that invalidates the reasoning and (2) the implementation may contain a programming error. (Recall the discussion in the introduction to Chapter 3.)

The following procedure can be used to check the final values of a and b for consistency with Specification 5.1:

Test 5.1
    .. Score *a* and *b* produced by a root refining program.

    .. 0 = failure, while 1 and 2 signal success.

    if *a, b* is not a sign-change of *f*

        score ← 0   ..no root is bracketed

    else if *ulps(a, b)* ≤ 1

        score ← 1

    else if ($a = 0$ and $|b| < 2\sigma$) or ($b = 0$ and $|a| < 2\sigma$)

        score ← 2

    else

        score ← 0   .. *a* and *b* are too far apart

For floating-point systems that do not satisfy the assumptions of Section 2.1, or for nonbinary machines, the accuracy of Algorithm 5.1 may be diminished. For instance, on a three-digit decimal ($\beta = 10$) machine, the computed midpoint of 0.596 and 0.600 is 0.600, showing that on such a machine Algorithm 5.1 could terminate with a and b several ulps apart. (Computing the midpoint as $a + (b - a)/2$ is better in this regard but suffers from potential problems with underflow.)

The number of function evaluations required by Algorithm 5.1 is easy to estimate. Each bisection halves the uncertainty, that is, gains an additional bit of accuracy, at the cost of a single function evaluation. Since $10 \approx 2^{3.3}$, roughly 3.3 iterations are needed to gain an extra decimal digit. (Of course, Algorithm 5.1 is not guaranteed to converge to a root of the user's intended function f. The claim is that it converges to a sign-change of whatever function is actually computed. Thus we have neatly side-stepped questions about the effect on the computed root of rounding errors in the procedure that evaluates f.)

For example, consider repeated bisection of an initial interval [1,2] on a machine with 24-bit floating-point fractions. For any f, Algorithm 5.1 requires 24 function samples to reach full machine accuracy. The main deviation from this sort of uniform behavior comes in cases where a and b converge to zero.

For example, consider a function f that satisfies $f(x) < 0$ for $x \leq 0$ and $f(x) > 0$ for $x > 0$. If the underflow threshold is $\sigma = 2^{-129}$ and if initially $[a, b] = [-1, 1]$, then 131 function evaluations (at $-1, 0, 2^{-1}, 2^{-2}, \ldots, 2^{-129}$) are made before Algorithm 5.1 terminates. See Exercise 4 of Section 2.1 for more details.

EXERCISES

1. Let $f(x) = 9x - 1$ and consider a floating-point number system with $\beta = 10$ and $p = 3$. Show that the computed value of f is never zero.
2. (a) Show that the following test occasionally fails to correctly determine whether a, b is a sign-change of f:

    if $f(a) = 0$ or $f(b) = 0$

        if $a = b$

            return *true*

    else

        if $a < b$ and $f(a) \times f(b) < 0$

            return *true*

    return *false*

   (b) Show that the above test for a sign-change will work correctly if the expression $f(a) \times f(b)$ is replaced by $(f(a)/|f(a)|) \times f(b)$.
3. Show that Algorithm 5.1 eventually terminates under the assumptions that (1) the variables a, b, c can only assume values from a finite set of "machine numbers" and that (2) comparisons, like $\leq$, between machine numbers work as expected. Show that additional assumptions are needed to prove termination if the stopping criterion
    $$c \leq a \text{ or } c \geq b$$
    is replaced by
    $$c = a \text{ or } c = b$$
4. Verify condition B.
5. Suppose that $\beta \geq 4$ and let a and b be floating-point numbers that satisfy $1 \leq a \leq b \leq 2$. Show that the computed value of $(a + b)/2$ is in error by at most half an ulp. Why does this imply that an implementation of Algorithm 5.1 should pass Test 5.1 on problems where initially $a = 1$ and $b = 2$ *even on nonbinary machines*?

5.2 INCORPORATING LINEAR INTERPOLATION

This section explains a modification to Algorithm 5.1 that increases its efficiency for many functions f. The improvement can be made without disastrously degrading worst-case performance and without sacrificing satisfaction of Specification 5.1.

The basic idea is to begin with the working assumption that f is well-behaved on the interval $[a, b]$. For the simple algorithm explained below we will assume that

1. f is nearly linear between a and b and that
2. No straight line intersects the graph of f at more than two points between $x = a$ and $x = b$; in other words, the graph is either concave between a and b or convex between a and b.

[The second condition is true in theory if a and b closely straddle a root ρ where the second derivative $f''(\rho)$ is nonzero.] If these working assumptions later appear incorrect or if they do not speed convergence, then we will perform a bisection step and try the assumptions on the reduced interval.

Let c be the intersection of the x-axis with the line L that interpolates f at a and b. See Figure 5.2. The slope of L can be expressed as either $(f(b) - f(a))/(b - a)$ or $(-f(a))/(c - a)$ (since the slope of a line is defined as the ratio of vertical change to horizontal change). Equating these two expressions and solving for c gives

$$c = a - f(a) \times \frac{b - a}{f(b) - f(a)}$$

Our next step depends on the sign of $f(c)$.

First suppose that $sign(f(a)) = sign(f(c))$. Let d be the root of the line L' interpolating f at a and c. Assumption (2) implies that between $x = c$ and $x = d$ the graph of f lies between L and L', so $sign(f(a)) \neq sign(f(d))$. Thus we can update $a \leftarrow c$ and, if assumption (2) is valid, $b \leftarrow d$. If, contrary to our expectations, $sign(f(a)) = sign(f(d))$ (this can happen if f has both concave and convex segments between a and b), we can raise the value of a still further by updating $a \leftarrow d$. See Figure 5.3.

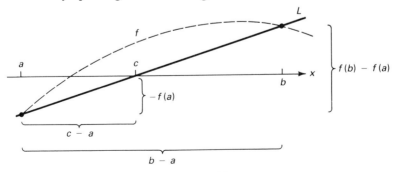

Figure 5.2

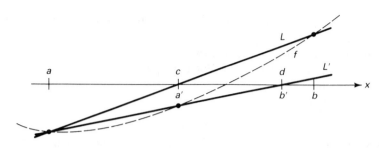

Figure 5.3

Now suppose that $sign(f(a)) \neq sign(f(c))$. Reasoning as before, we are led to expect, based on assumption (2), that if d is the root of the line L'' that interpolates f at c and b, then $sign(f(a)) = sign(f(d))$. Thus we can update $b \leftarrow c$ and (probably) $a \leftarrow d$. See Figure 5.4.

If anything goes wrong with the above process, for instance, if $d \leq a$ or $d \geq b$, then we will revert to bisection. Moreover, if the step is completed but does not at least halve the interval length $b - a$, then we will perform a bisection step. Samples $f(x)$ will be taken only for x between the current values of a and b, and, regardless of the outcome, each such function evaluation will allow us to either raise a or lower b.

Since Algorithm 5.2 will never take more than three function evaluations to halve $b - a$, the performance of Algorithm 5.1 has not been disastrously degraded. Each additional bit of accuracy now costs at most three function evaluations. Moreover, no accuracy has been lost since Algorithm 5.2 terminates under the same conditions as Algorithm 5.1.

Implementation Blunders. There are at least three distinct reasons why testing may be relatively ineffective at exposing mistakes in an implementation of Algorithm 5.2:

1. A mistake in the interpolation phase is liable to result in the appearance of symptoms that interpolation is not converging fast enough and thereby cause the program to revert to bisection. For example, a mistake in the expression for c or for d would slow convergence but would not affect the correctness of the final results.

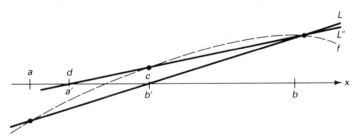

Figure 5.4

Algorithm 5.2

    .. If at any time a computed value $f(x)$ is zero,
    .. set $a \leftarrow x, b \leftarrow x$, and return.
interpolate:
        $length \leftarrow b - a$
        $c \leftarrow a - f(a) \times \dfrac{b - a}{f(b) - f(a)}$
        if $c \leq a$ or $c \geq b$
                go to bisection
        if $sign(f(a)) = sign(f(c))$
                $d \leftarrow a - f(a) \times \dfrac{c - a}{f(c) - f(a)}$
                $a \leftarrow c$
        else
                $d \leftarrow b - f(b) \times \dfrac{c - b}{f(c) - f(b)}$
                $b \leftarrow c$
        if $d \leq a$ or $d \geq b$
                go to bisection
        if $sign(f(a)) = sign(f(d))$
                $a \leftarrow d$
        else
                $b \leftarrow d$
        if $b - a \leq length / 2$
                go to interpolate
bisection:
        $c \leftarrow \dfrac{a + b}{2}$
        if $c \leq a$ or $c \geq b$
                return
        if $sign(f(a)) = sign(f(c))$
                $a \leftarrow c$
        else
                $b \leftarrow c$
        go to interpolate

2. An acceptable implementation of Algorithm 5.2 will contain logical expressions that are almost never true and instructions that are rarely executed. For instance, unless the testing process is specifically designed to check all code that handles $f(x) = 0$ or that guards against division by zero, then mistakes in those statements will probably go undetected.
3. When interpolation is converging properly, one iteration will reduce *length* $= b - a$ far more than by half. This means that the process is unaffected by many potential mistakes in computing *length* and in the decision whether to repeat the interpolation step.

These observations suggest the following experimental hypothesis:

Hypothesis 5.1
Test 5.1 is ineffective at exposing implementation blunders in Algorithm 5.2.

We performed a mutation experiment on (1) a straightforward implementation of Algorithm 5.2, (2) the set of data $f(x) = s^3 + s$, where $s = x - 1.4$, $a = 1$, $b = 2$, and the similar set of data with 1.4 replaced by 1.6 (see Figure 5.5), and (3) Test 5.1 as the acceptance criterion. These two sets of data caused execution of all branches of the three *if-else* statements in the above listing of Algorithm 5.2. (In other words, every assignment statement in that listing was executed at least once during the solution of these two problems.)

Of the 1365 possible typographical changes, 761 (about 56.8%) produced a mutant that performed a valid computation for both sets of data and whose computed solutions satisfied Test 5.1. An approach that diagnoses many of the remaining mistakes is discussed in the next section.

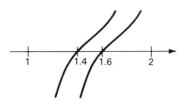

Figure 5.5

PROGRAMMING ASSIGNMENT

Implement Algorithm 5.2. Your procedure should never evaluate f twice at the same argument. If a value of f equal to zero is found, then no further samples should be taken. Also, division by zero should not be possible. [*Hint:* Under what conditions will the tests $f(c) - f(a) = 0$ and $f(c) = f(a)$ give different results?] Test your program on at least one problem where it should score 1 on Test 5.1 and one problem on which it should score 2. [*Hint*: Consider the function defined by the following: if $x = 0$ then $f \leftarrow \sigma$ else $f \leftarrow x$.]

EXERCISE

1. This exercise investigates the accuracy of the interpolation formula $c = a - \delta$, where $\delta = f(a) \times [(b - a)/(f(b) - f(a))]$. The values a, b, $f(a)$, and $f(b)$ are to be considered exact.
   (a) Show that δ is computed to within a few ulps if underflow and overflow do not occur.
   (b) Assume that $sign(f(a)) \neq sign(f(b))$ and that $0 < a < b$. Show that c is computed to within a few ulps unless underflow or overflow occur.
   (c) Show that c is substantially more susceptible to underflow than is the midpoint expression $(a + b)/2$ when a and b have the same sign.

5.3 THE IMPORTANCE OF PERFORMANCE MEASUREMENTS

The process of performance measurement is less clear-cut than is testing (as the terms were defined in Section 1.1). Testing presupposes a precise statement of what to look for, namely the specification; measuring is done without a precise idea of what counts as acceptable program behavior. In brief, the outcome of a test is a single bit, *accept* or *reject*, while performance measurements are much harder to summarize.

However, performance measurements are more than merely descriptive; they often lead directly to changes in the program. That is, we do not just passively measure a program's performance characteristics, as we would with a person's height. (Even if we think that the person's height is not desirable, we will not try to change it.) Performance measurements are best conducted with as concrete an idea as one can get, based on intuition or experience, of what behavior should be regarded as satisfactory. Much of the code one finds in good numerical software can be directly attributed to unacceptable measured performance of an earlier, and simpler, version of that software.

This section contains two illustrations of the importance of performance measurements in the development of root refining programs. First we will describe a mutation experiment that supports the following hypothesis:

Hypothesis 5.2
Performance measurements are able to expose a significant proportion of the implementation blunders in Algorithm 5.2 that are not revealed by Test 5.1.

Second, we will show that systematic performance measurements can lead directly to substantial improvements in the efficiency of Algorithm 5.2.

5.3.1 A Mutation Experiment Supporting Hypothesis 5.2

Of the three reasons, given in Section 5.2, why Test 5.1 is relatively ineffective at exposing mistakes in an implementation of Algorithm 5.2, the first suggests that a number of potential mistakes would be revealed by inefficiency of the program. On the other hand, the remaining reasons point to classes of mistakes that make a difference in program performance only under special circumstances.

We repeated the experiment described at the end of the previous section but with the added acceptance criterion that the program could use no more than eight samples of f. (Our single precision implementation of Algorithm 5.2 required eight function values for each of the two problems used for the experiment.) Of the 761 program mutants that passed Test 5.1, 387 (just over half) were exposed because they invoked one of the two functions f more than eight times.

5.3.2 Systematic Measurement of Root Refining Programs

It is necessary to measure the efficiency of an implementation of Algorithm 5.2, since we have no particularly useful theoretical results about its rate of convergence on smooth functions. Here we will describe the use of a collection of root refining problems, each of which begins with an initial interval $a = 1$, $b = 2$. (Exercise 5 of Section 5.1 explains why this choice may be desirable for an initial interval.) The problems are divided into four groups; each group consists of polynomials of degree 3, 5, 7, and 9. For each group and each choice of polynomial degree, we will use 10 functions $f_\alpha(x)$ that depend on a randomly chosen parameter α. Thus, we will measure the performance of a root refining program on 160 sets (f, a, b) of data.

Group 1. $f(x) = s^n$, where $s = x - \alpha$ and $n = 3, 5, 7, 9$. See Figure 5.6. Lines interpolating f around the solution α are nearly horizontal, and hence their intersections with the x-axis are extremely sensitive to errors. We can expect these test problems to be relatively difficult for Algorithm 5.2. (On the other hand, these problems are easier for the bisection method of Algorithm 5.1 than are the problems in the other three groups. See Exercise 1 below.)

Group 2. $f(x) = s^n + 10^{-4}$, where $s = x - \alpha$ and $n = 3, 5, 7, 9$. The solution is $\alpha - 10^{-4/n}$ which must exceed 1 if we pick $\alpha > 1.5$. See Figure 5.7. It appears that a fair number of bisections may be needed to reach an interval in which the assumption of near-linearity is valid.

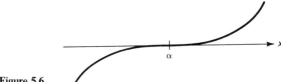

Figure 5.6

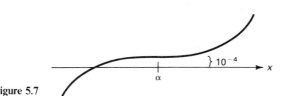

Figure 5.7

Group 3. $f(x) = s^n + s$, where $s = x - \alpha$ and $n = 3, 5, 7, 9$. Each of the four problems has solution α, and each f crosses the x-axis at a 45-degree angle. See Figure 5.8. These functions are quite close to being linear around $x = \alpha$, but they violate Algorithm 5.2's motivating assumption that

no straight line intersects the graph of f at more than two points (†)

In particular, any straight line that crosses the x-axis at $x = \alpha$ and has slope greater than 1 meets the graph of f at three points.

Group 4. $f(x) = s^n + s + 10^{-4}$, where $s = x - \alpha$ and $n = 3, 5, 7, 9$. The solution is close to $\alpha - 10^{-4} + 10^{-4n}$. See Figure 5.9. There seems to be no reason to expect difficulty with these problems. Even if assumption (†) turns out to be vital to the efficiency of Algorithm 5.2, 13 bisections are adequate to shrink $[a, b]$ far enough that it excludes α, which makes assumption (†) valid.

The following FORTRAN programs exercise a double precision root refining procedure called *ZERO*. For each group of problems and each choice of polynomial degree, the program computes the lowest, highest, and average number of function evaluations for 10 functions $f_\alpha(x)$. The f_α are determined using a double precision random number generator *RAN* to select an α

Figure 5.8   Figure 5.9

satisfying $1.5 < \alpha < 2.0$. [This condition on α guarantees that $f(1) < 0 < f(2)$.] Note how the FORTRAN *COMMON* statement is used to communicate values between the main procedure and the function *F*.

```
      COMMON ALPHA, LINEAR, CONST, N, NCALLS
      EXTERNAL F
      DOUBLE PRECISION F, ALPHA, A, B, RAN
      INTEGER LINEAR, CONST, N, NCALLS, LO, HI, TOTAL,
     *TIMES
      PRINT * ,'            N   LO   HI  AVERAGE'
      DO 30 LINEAR = 0, 1
      DO 30 CONST = 0, 1
          PRINT 1, LINEAR, CONST*0.0001
1         FORMAT( / 'S**N + ',I1,'*S + ', F5.4)
          DO 20 N = 3,9,2
              LO = 1000
              HI = 0
              TOTAL = 0
              DO 10 TIMES = 1,10
                  ALPHA = 0.5D0*RAN( ) + 1.5D0
                  A = 1.0D0
                  B = 2.0D0
                  NCALLS = 0
                  CALL ZERO(F,A,B)
                  LO = MIN(LO,NCALLS)
                  HI = MAX(HI,NCALLS)
                  TOTAL = TOTAL + NCALLS
10            CONTINUE
              PRINT 11, N, LO, HI, TOTAL / 10.0
11            FORMAT(15X,3I6,F6.1)
20        CONTINUE
30    CONTINUE
      END

      DOUBLE PRECISION FUNCTION F(X)
      COMMON ALPHA, LINEAR, CONST, N, NCALLS
      DOUBLE PRECISION X, S, ALPHA
      INTEGER LINEAR, CONST, N, NCALLS
      NCALLS = NCALLS + 1
      S = X - ALPHA
      F = S**N + LINEAR*S + CONST*0.0001D0
      RETURN
      END
```

Sec. 5.3 The Importance of Performance Measurements

Here are the results of conducting such a performance measurement on a machine with 56-bit double precision fractions. The wide range among the measurements for the second problem group vindicates our decision to sample several polynomials of each type.

	N	LO	HI	AVERAGE
S**N + 0*S + .0000				
	3	75	86	81.4
	5	55	58	56.3
	7	43	46	44.4
	9	30	37	33.9
S**N + 0*S + .0001				
	3	20	47	31.5
	5	18	43	33.1
	7	15	42	27.8
	9	16	39	31.3
S**N + 1*S + .0000				
	3	10	10	10.0
	5	7	7	7.0
	7	7	7	7.0
	9	7	7	7.0
S**N + 1*S + .0001				
	3	19	35	31.9
	5	60	60	60.0
	7	60	60	60.0
	9	60	60	60.0

The biggest surprise concerns the last group of problems. Computing the root of $s^5 + s + 10^{-4}$ involves 60 function evaluations, more than the 56 required by bisection. What went wrong?

A little detective work, carried out by instrumenting the program to display intermediate computed values, turns up the following fact. After only a few evaluations of f, one of the end points a or b can reach its final value, and each linear interpolation thereafter produces $c = a$ or $c = b$.

5.3.3 Improving the Performance of Our Root Refining Program

Suppose that a is so near to a root of f that the computed value $f(a)$ is less than the distance from a to the next floating-point number. See Figure

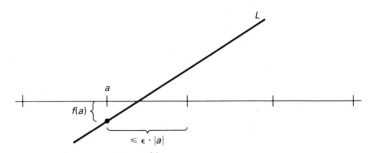

Figure 5.10

5.10. An attempt to find the root of a line L interpolating f at a and another point b is liable to compute a. In algebraic terms, if $(b - a)/(f(b) - f(a))$ (the inverse of the slope of L) has order of magnitude 1, then a may well be the closest floating-point number to $a - f(a) \times [(b - a)/(f(b) - f(a))]$.

We propose the following modification to Algorithm 5.2. When $c = a$ (where c is the computed root of the line L that interpolates f at a and b), then move c an ulp or so toward b before continuing. Similarly, move c an ulp or so toward a if the computed value of c is b. One possibility is the following:

Algorithm 5.3a
Modify Algorithm 5.2 as follows: After the statement

$$c \leftarrow a - f(a) \times \frac{b - a}{f(b) - f(a)}$$

insert the lines

  if $c \leq a$

    $c \leftarrow a + \epsilon \times |a|$

  if $c \geq b$

    $c \leftarrow b - \epsilon \times |b|$

where ϵ = machine epsilon.

Exercise 2 justifies these changes and suggests a modification that produces a slightly more efficient algorithm.

Here are the results of running an implementation of Algorithm 5.3a on the problems discussed above:

	N	LO	HI	AVERAGE
S**N + 0*S + .0000				
	3	80	95	90.0
	5	58	61	59.3
	7	45	48	46.5
	9	30	38	34.4

Sec. 5.3   The Importance of Performance Measurements   125

```
S**N + 0*S + .0001
              3    17    23   20.8
              5    14    22   19.1
              7    15    20   17.3
              9    13    18   16.4
S**N + 1*S + .0000
              3    10    10   10.0
              5     7     7    7.0
              7     7     7    7.0
              9     7     7    7.0
S**N + 1*S + .0001
              3    12    12   12.0
              5    10    10   10.0
              7    10    10   10.0
              9    10    10   10.0
```

Notice that efficiency has been almost doubled on group 2 of test problems and increased about sixfold on group 4.

5.3.4 The Moral of This Story

In this case, as is often true, performance measurement was especially useful because we formed some expectations about the outcome and *then* conducted the measurements. It is because we anticipated that the problems in group 4 would be "easy" that the measurements reported above point the way to a better program.

PROGRAMMING ASSIGNMENTS

1. Measure the performance of your implementation of Algorithm 5.2 using the program given in this section, modified to include Test 5.1.

2. Implement Algorithm 5.3a with a modification that improves its performance on the problems in group 1. [*Hint*: Perform a bisection whenever the slope S of the interpolating line satisfies $|S| \leq \Theta$. Desirable values for the threshold Θ can only be found by trial and error. For instance, you might try a few values between 0.01 and 0.0001 and pick the one that works best for the problems in groups 1–4. Much of the best numerical software contains such "magic constants" whose only justification is that they seem to work.]

3. (Optional) Implement Algorithm 5.3b, which is based on "Two Efficient Algorithms with Guaranteed Convergence for Finding a Zero of a Function" by J. C. P. Bus and T. J. Dekker (*ACM Transactions on Mathematical Software*, Dec. 1975, pp. 330–345). [*Warning*: Algorithm 5.3b is surprisingly difficult to comprehend. In the paper "Understanding and Documenting Programs" (*IEEE Transactions on Soft-*

ware Engineering, May 1982, p. 282), Victor Basili and Harlan Mills hazard the guess that it would take several weeks for a maintenance programmer versed in their approach to develop and document an understanding of a related program. However, Algorithm 5.3b is given in enough detail that it can be implemented without being understood.]

Algorithm 5.3b

$\epsilon \leftarrow$ machine epsilon

fa $\leftarrow$ *f*(*a*)

if *fa* = 0.0

    b $\leftarrow$ *a*; return

fb $\leftarrow$ *f*(*b*)

if *fb* = 0.0

    a $\leftarrow$ *b*; return

first $\leftarrow$ *true*

interpolate:

    c $\leftarrow$ *a*; *fc* $\leftarrow$ *fa*; *ext* $\leftarrow$ 0

extrapolate:

    if |*fa*| < |*fb*|　　　　　..interchange

        if *c* ≠ *a*

            d $\leftarrow$ *a*; *fd* $\leftarrow$ *fa*

        a $\leftarrow$ *b*; *fa* $\leftarrow$ *fb*; *b* $\leftarrow$ *c*; *fb* $\leftarrow$ *fc*; *c* $\leftarrow$ *a*; *fc* $\leftarrow$ *fa*

    $m \leftarrow \dfrac{c + b}{2}$

    if $m \leq \min(c, b)$ or $m \geq \max(c, b)$

        a $\leftarrow$ *c*; return

    mb $\leftarrow$ *m* − *b*

    if *ext* > 3

        w $\leftarrow$ *mb*

    else

        p $\leftarrow$ (*b* − *a*) × *fb*

        if (*first*)

            q $\leftarrow$ *fa* − *fb*; *first* $\leftarrow$ *false*

Sec. 5.3 The Importance of Performance Measurements

else

$$fdb \leftarrow \frac{fd - fb}{d - b}; \quad fda \leftarrow \frac{fd - fa}{d - a}; \quad p \leftarrow fda \times p;$$
$$q \leftarrow fdb \times fa - fda \times fb$$

if $p < 0.0$

$p \leftarrow -p; \quad q \leftarrow -q$

if $ext = 3$

$p \leftarrow 2p$

$least \leftarrow 0.6 \times \epsilon \times |b| \times sign(mb)$

if $p = 0.0$ or $p \leq q \times least$

$w \leftarrow least$

else if $p < mb \times q$

$w \leftarrow p / q$

else

$w \leftarrow mb$

$d \leftarrow a; \quad fd \leftarrow fa; \quad a \leftarrow b; \quad fa \leftarrow fb; \quad b \leftarrow b + w; \quad fb \leftarrow f(b)$

if $fb = 0.0$

$a \leftarrow b$; return

if $sign(fc) = sign(fb)$

go to interpolate

if $w = mb$

$ext \leftarrow 0$

else

$ext \leftarrow ext + 1$

go to extrapolate

Use performance measurements to investigate the following claims: *For functions f that are tangent to the x-axis, the solution to Programming Assignment 2 is usually several times more efficient than the implementation of Algorithm 5.3b. For other functions f, Algorithm 5.3b is usually the more efficient of the two by a margin of 10%–30%.*

Are there additional "features" of the data (f, a, b), other than whether or not f's slope at the root is zero, that are useful for predicting which program will be more efficient?

EXERCISES

1. Suppose that Algorithm 5.1 is applied with initial values $a = 1$ and $b = 2$ on a machine with $\beta = 10$, $p = 16$, and $m = -80$. Show that in general about $3.3 \times 16 \approx 53$ function evaluations will be needed. Also show that at most $3.3 \times 9 \approx 30$ function evaluations will be needed for the special case of $f(x) = (x - \alpha)^9$, provided that Algorithm 5.1 is implemented so that it terminates if it finds an x with $f(x) = 0$. [*Hint*: Underflow occurs when $|c - \alpha| < 10^{-9}$.]
2. Consider a machine with floating-point base $\beta = 2$, and let ϵ denote machine epsilon.
   (a) Argue that for about half of all floating-point numbers a, the computed value of $a + \epsilon \times |a|$ is one representable number to the right of a, while for the other half, it is two representable numbers away.
   (b) Explain why the computed value of $a + 0.6 \times \epsilon \times |a|$ is almost always the floating-point number immediately to the right of a. When does this not hold? What if $\beta > 2$?

5.4 MINIMIZATION (OPTIONAL)

Many computational problems that include a function f among their data cannot be solved by a foolproof numerical method. Typically, the approach taken to these problems is to develop algorithms whose behavior can be analyzed assuming that the data meet certain assumptions and that arithmetic is exact. The hope is that the analysis is indicative of actual behavior in cases where the machine's precision far exceeds the desired accuracy of the computed solution.

In this section we will illustrate the following potential obstructions to this approach:

1. In theory, these analyses of algorithm behavior are usually irrelevant in the realm of inexact arithmetic because they rest on assumptions that rarely hold, or make no sense, for floating-point number systems.
2. In practice, these analyses are sometimes quite misleading because of their failure to take computational errors into account.

The algorithms given to illustrate these points have the following properties, common in programs whose data includes a function f:

a. The failure of the exact-arithmetic analysis to predict program performance is easy to discover by empirical means. This is in contrast to Example 1 of Section 2.4 and the results of Experiment 4.1a, which deal with "algebraic" problems for which numerical instabilities are hard to uncover through testing.
b. Actual program performance is correctly predicted by an analysis that requires only a vague notion of computational error; no detailed model

Sec. 5.4 Minimization (Optional)

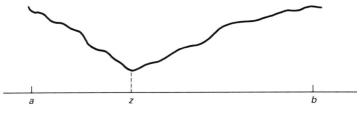

Figure 5.11

of floating-point arithmetic is needed. Recall that the situation is quite different for sine procedures. For instance, Exercise 1 of Section 3.4 shows that a seemingly minor relaxation of the assumptions stated in Section 2.1 (namely, the absence of a guard digit for subtraction) has significant repercussions.

The Minimization Problem. Given a function f and numbers a and b, with $a < b$, the problem is to find a number z between a and b that minimizes f. In other words, z is to have the properties (1) $a \leq z \leq b$ and (2) if x is any number such that $a \leq x \leq b$, then $f(z) \leq f(x)$. Although minimization is in some respects closely analogous to solving nonlinear equations, we are particularly interested in ways in which the analogy breaks down.

It is a theorem of mathematics that the minimization problem has a solution z provided that f is a continuous function. However, there is no hope of writing a program that is guaranteed to find z given an arbitrary continuous f. See Exercise 1.

Analysis of an Algorithm. We will develop an algorithm that is guaranteed to solve the minimization problem whenever the graph of f slopes downward between a and z and slopes upward between z and b. See Figure 5.11. More precisely, the requirement is that there exists a number z, with $a \leq z \leq b$, that possesses the following properties:

(i) If $a \leq x < y \leq z$, then $f(x) > f(y)$.
(ii) If $z \leq x < y \leq b$, then $f(x) < f(y)$.

Such functions are said to be *unimodal* between a and b. The functions shown in Figure 5.12 are *not* unimodal between a and b. It is possible to prove

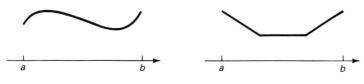

Figure 5.12

theorems whose intuitive thrust is that for almost all well-behaved functions f, (i) f is unimodal if we pick a and b sufficiently close to a number z that minimizes f and (ii) the algorithm given below converges to a value z that minimizes f over some (perhaps very small) range of values $a \le x \le b$.

We will now prove that there exists a numerical procedure that solves the minimization problem for all sets f, a, b of data having the property that f is unimodal between a and b. To do so, it is not necessary to pay much attention to issues of efficiency; we need only display an algorithm that works correctly. However, we will take some pains to present an algorithm that is actually rather useful. The procedure that we will develop is called the *golden section search* method.

Our construction mimics the approach followed in the bisection method. That is, we will sample f in the "search region" between a and b and, as soon as possible, discard a part of the search region that cannot contain the solution. The process can be repeated on the reduced search region.

Clearly, one sample of f is not enough to rule out any part of the search region. See Figure 5.13. However, two samples suffice. See Figure 5.14. The only trick is to sample f in a systematic way so that (1) at each iteration the sample of f in the interior of the reduced search region can be used again, meaning that only one new sample of f is needed to further reduce the region, and (2) the discarded portion of the search region is as large as possible.

Let us call the intervals from a to c and from d to b the *major intervals* and call the interval from c to d the *minor interval*. Our goal is to arrange the computation so that the minor interval becomes one of the major intervals for the next iteration, as indicated in Figure 5.15.

Let r denote the length of a major interval divided by the length of the search region, that is,

$$r = \frac{c - a}{b - a} = \frac{b - d}{b - a}$$

There is only one value of r that will work to make the spacing among a', c', d', and b' in the reduced search region exactly proportional to the spacing among a, c, d, and b, and our next task is to determine that value of r.

To simplify the formulas, suppose $a = 0$, $b = 1$, and $f(c) < f(d)$. (A value of r that works in general will work in this special case.) Then in Figure

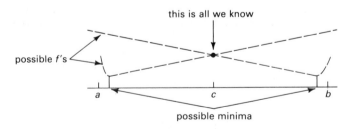

Figure 5.13

Sec. 5.4 Minimization (Optional)

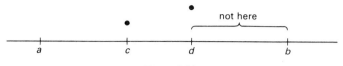

Figure 5.14

5.15 we have $c = r$, $d = 1 - r$, and $d - c = 1 - 2r$. But since the proportions are to remain the same in the reduced search region and since the interval from d' to b' is a major interval of the reduced region, we have

$$r = \frac{b' - d'}{b' - a'} = \frac{d - c}{d - a} = \frac{d - c}{d} = \frac{1 - 2r}{1 - r}$$

It follows, upon equating the first and last terms, that

$$r(1 - r) = 1 - 2r$$

or, equivalently,

$$r^2 - 3r + 1 = 0$$

This quadratic equation has two solutions r, but since we are only interested in values $r < 1$, we have

$$r = \frac{3 - \sqrt{5}}{2}$$

We can summarize these formulas as follows. If we let $r = (3 - \sqrt{5})/2$, then the major intervals of the search region both have length $M = r \times (b - a)$. If $f(c) < f(d)$, we can update $b \leftarrow d$, $d \leftarrow c$, and $c \leftarrow a + r \times (b - a)$. In addition, since $f(d)$ is just the previous value of $f(c)$, only $f(c)$ need be computed to repeat the process. A similar approach can be followed in the case $f(c) \geq f(d)$. This leads to the following algorithm.

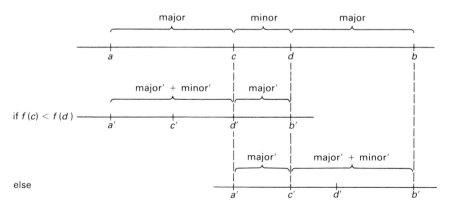

Figure 5.15

Algorithm 5.4a

$$r \leftarrow \frac{3 - \sqrt{5}}{2}$$

$M \leftarrow r \times (b - a)$

$c \leftarrow a + M$

$d \leftarrow b - M$

repeat until satisfied

    if $f(c) < f(d)$

        $b \leftarrow d$

        $d \leftarrow c$

        $c \leftarrow a + r \times (b - a)$

    else

        $a \leftarrow c$

        $c \leftarrow d$

        $d \leftarrow b - r \times (b - a)$

Since the distance from a to c equals that from d to b, we have $c = a + (c - a) = a + (b - d)$. This observation leads to the following variant of golden section search, which avoids a multiplication at each iteration:

Algorithm 5.4b

$$r \leftarrow \frac{3 - \sqrt{5}}{2}$$

$M \leftarrow r \times (b - a)$

$c \leftarrow a + M$

$d \leftarrow b - M$

repeat until satisfied

    if $f(c) < f(d)$

        $b \leftarrow d$

        $d \leftarrow c$

        $c \leftarrow a + (b - d)$

    else

        $a \leftarrow c$

        $c \leftarrow d$

        $d \leftarrow b - (c - a)$

Sec. 5.4 Minimization (Optional)

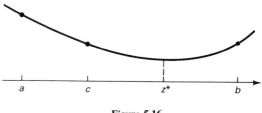

Figure 5.16

Notice that since $r \approx 0.3820$, the length $b' - a'$ of the reduced search region is $1 - r$ (≈ 0.618) times the length $b - a$ of the original region. In other words, once the process gets going, each function evaluation allows us to reduce the size of the search region by a factor of around 0.618. Although this is not as rapid as the reduction obtained with the bisection process for finding roots, three function evaluations in the golden section procedure are better than two function evaluations with bisection. This holds because three golden section steps reduce the search region by the factor $(0.618)^3 \approx 0.236 < 0.25 = (0.5)^2$, the reduction gained by two bisections.

As with bisection, it is possible to blend the slow but steady method with a procedure that converges much faster when f is smooth. In this capacity, the role assumed in Section 5.2 by linear interpolation can be played by "parabolic" interpolation. That is, given $f(a)$, $f(b)$, and $f(c)$, we can approximate the solution z by the value z^* that minimizes the interpolating polynomial of degree 2. See Figure 5.16. We will not pursue this avenue further; the interested reader can consult *Algorithms for Minimization Without Derivatives* by Richard Brent (Prentice-Hall, Englewood Cliffs, N.J., 1973).

The following table summarizes the analogies we have drawn between root finding and minimization:

	Root Finding	Minimization
Solvable case	$sign(f(a)) \neq sign(f(b))$	f is unimodal
Reliable method	Bisection	Golden section search
Efficient method	Linear interpolation	Parabolic interpolation

Now we turn to some differences between the two problems.

Irrelevance of the Analysis. The notion that "f is unimodal between a and b" makes sense in the context of inexact arithmetic. We need only think of the values a, b, x, y, z, $f(x)$, and $f(y)$ that appear in the definition on page 129 as belonging to a floating-point number system.

However, the notion is essentially useless in this context because few functions are unimodal as they are actually computed. For instance, consider

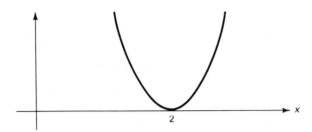

Figure 5.17

the function

$$f(x) = 1.5x^2 - 6.0x + 6.0$$

which is unimodal in exact arithmetic. See Figure 5.17. Suppose that f is evaluated as $f \leftarrow (1.5 \times x - 6.0) \times x + 6.0$ using three-digit decimal arithmetic (with the round-to-even rule for breaking ties). Thus $f(2.03)$ would be computed as follows:

$$1.5 \times 2.03 = 3.045 \rightarrow 3.04$$
$$3.04 - 6.0 = -2.96$$
$$-2.96 \times 2.03 = -6.0088 \rightarrow -6.01$$
$$-6.01 + 6.0 = -0.01$$

The graph (Figure 5.18) of the computed values of f for the 21 floating-point numbers between 1.9 and 2.1 is far from unimodal. Thus, strictly speaking, the notion of a function being unimodal rarely applies in the realm of fixed-precision arithmetic.

Inconsistency of the Analysis with Practice. The above analysis accurately predicts actual program performance of Algorithm 5.4a. In particular, it correctly predicts that minimizing $f(x) = (x - \sqrt{2})^2$, from an initial interval of $a = 1$ and $b = 2$, will produce $\sqrt{2}$ to nearly full machine accuracy at a cost of just under $1.5 \times p$ function values, where p is the number of bits in a floating-point fraction. (That is, golden section search needs about 50% more function evaluations than does the bisection procedure.) On the other hand, the analysis is quite misleading for Algorithm 5.4b. A more realistic

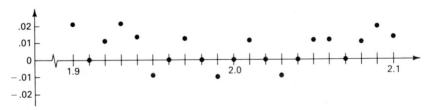

Figure 5.18

Sec. 5.4 Minimization (Optional)

analysis indicates that Algorithm 5.4b is incapable of computing more than about half the figures of the solution. See Programming Assignment 1.

The suggestions given in Exercise 4 will help you perform an analysis that is adequate for predicting the impact of computational errors on Algorithm 5.4b. If you carry out the analysis, notice that it does not require a detailed model of floating-point arithmetic.

PROGRAMMING ASSIGNMENTS

1. Use implementations of Algorithms 5.4a and 5.4b in an attempt to compute $\sqrt{2}$ to approximately full machine accuracy by minimizing $f \leftarrow (x - \sqrt{2})^2$.

2. (Optional) Design and implement a root bracketing procedure. [*Hint:* Suppose $f(0) > 0$. Try to find numbers a, c, and b such that $a < c < b$, $f(a) > f(c)$, and $f(c) < f(b)$. Then perform golden section search between a and b.]

   Make the following measurements of the procedure's reliability and efficiency. Generate 100 random numbers α satisfying $-2 \leq \alpha \leq 2$. For each α, let $f_\alpha(x)$ be the function computed by

   $$t \leftarrow 50 \times (x - \alpha)$$
   $$f \leftarrow 1 - 2/(1 + t^2)$$

   See Figure 5.19. Record (1) the number of times that the procedure successfully brackets a root and (2) the average number of function evaluations.

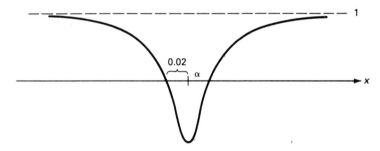

Figure 5.19

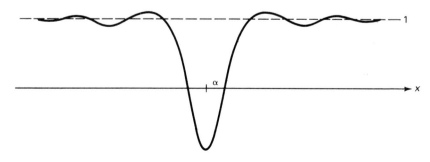

Figure 5.20

In addition, make the same measurements for the 100 functions $g_\alpha(x)$ computed by

$$t \leftarrow 50 \times (x - \alpha)$$
$$g \leftarrow 1 - 2 \times \cos(t)/(1 + t^2)$$

See Figure 5.20. (The tables in Section 1.3 report these measurements for two simple-minded root bracketing procedures.)

EXERCISES

1. Show that the minimization problem is in general unsolvable. [*Hint*: Adapt the argument given for the root bracketing problem in Section 1.3.]
2. Show that Algorithm 5.4a is largely unaffected by the use of inexact arithmetic. In particular, argue informally that the *computed* values maintain the relationship $a < c < d < b$ until they are a few ulps apart, unless underflow occurs.
3. Show that the function $f \leftarrow (x - \sqrt{2})^2$ of Programming Assignment 1 is unimodal between 1 and 2. Include consideration of rounding errors and underflow.
4. (Optional and difficult) This exercise asks you to analyze Algorithm 5.4b under the assumptions that the computed value of M is incorrect by an amount e and that all subsequent operations are exact for their given operands. Let a_n, b_n, c_n, and d_n denote the true values of a, b, c, and d at the start of the nth iteration of the *repeat* loop; let $a'_n = a_n + \alpha_n$, $b'_n = b_n + \beta_n$, $c'_n = c_n + \gamma_n$, and $d'_n = d_n + \delta_n$ denote the corresponding computed values.
   (a) Show that if $f(c'_n) < f(d'_n)$, then the errors at the next iteration can be expressed in terms of the current errors as follows: $\alpha_{n+1} = \alpha_n$, $\beta_{n+1} = \delta_n$, $\gamma_{n+1} = \alpha_n - \gamma_n + \delta_n$, and $\delta_{n+1} = \gamma_n$. Derive similar expressions for α_{n+1}, β_{n+1}, γ_{n+1}, and δ_{n+1} in terms of α_n, β_n, γ_n, and δ_n for the case $f(c'_n) \geq f(d'_n)$. Let λ_n denote the difference between the length of the computed n^{th} minor interval and the true interval's length, i.e., $\lambda_n = (d'_n - c'_n) - (d_n - c_n) = (d'_n - d_n) - (c'_n - c_n) = \delta_n - \gamma_n$.
   (b) Show that if $\lambda_n < -(d_n - c_n)$, then $c'_n > d'_n$. Also show that if f is unimodal and $a'_n < d'_n < c'_n < b'_n$, then the value between a'_n and b'_n that minimizes f does *not* lie between a'_{n+1} and b'_{n+1}.
   (c) Under the above assumptions about computational error, show that $\lambda_1 = -2e$, $\lambda_2 = 3e$, and $\lambda_{n+2} = -\lambda_{n+1} + \lambda_n$ for $n \geq 1$. [*Hint*: By definition $\alpha_1 = \beta_1 = 0$. Use part (a) and consider all four combinations of possibilities for iterations n and $n + 1$.]
   It follows from the theory of linear difference equations that λ_n is closely approximated by $[(5 + 3\sqrt{5})/10]\rho^n e$, where $\rho = (-1 - \sqrt{5})/2$. (You might use a calculator to check this claim.)
   (d) Show that $\rho = -(1 - r)^{-1}$.
   The length of the true minor interval is $d_n - c_n = (b - a)(1 - 2r)(1 - r)^{n-1}$ since each iteration shrinks the initial length $d_1 - c_1 = (b - a)(1 - 2r)$ (where a and b denote the original end points) by an additional factor of $1 - r$. According to part (b), convergence ceases when this length falls below $|\lambda_n|$, i.e., dropping a few factors near 1.0, when

$$(b - a)(1 - r)^n \leq |\rho^n e|$$

Sec. 5.4 Minimization (Optional)

(e) Use part (d) to show that this condition is equivalent to
$$(b-a)(1-r)^n \le \sqrt{|e|(b-a)}$$
If ϵ = machine epsilon, then the inevitable errors in computing M leave it wrong by an amount e, with $|e| \approx \epsilon(b-a)$, so convergence ceases when $d_n - c_n \approx \sqrt{\epsilon}(b-a)$.

(f) Suppose that $a = 1 < c_n < d_n < 2 = b$ and that $d_n - c_n = \sqrt{\epsilon}(b-a)$. Show that d_n and c_n agree to only about half of the figures carried by the machine. Compare this observation with the results you obtained for Programming Assignment 1.

6

INTEGRATION

The purpose of this chapter is to investigate the process of measuring the performance of numerical software. The chapter can be summarized by the following remark, attributed to an anonymous numerical analyst: *It is an order of magnitude easier to write two sophisticated subroutines than to determine which of the two is better.*

Sections 6.1 and 6.2 develop a simple procedure that attempts to solve the following formulation of the problem of numerical integration:

The Problem of Automatic Integration
Let f, a, b, and tol be given, where f is a function (implemented by a specific computational procedure) and a, b, and tol are numbers that satisfy $a < b$ and $tol > 0$. Compute a number within tol of $\int_a^b f(x)\,dx$.

The remainder of the chapter treats, in the context of automatic integration procedures, two questions that are central to this book: *What claims can we make about the software we write? How can we gather evidence to support those claims?*

Section 6.3 indicates that our claims about automatic integration procedures must be limited in scope. In particular, claims that one integration procedure is more efficient than another will apply only to particular classes of integration problems, since it will generally be possible to find another class of problems where the second procedure outperforms the first. We will also see that our claims must be probabilistic in nature to allow for occasional algorithmic failures. Thus we will consider claims of the following form: *For data (f, a, b, tol) with the features..., procedure A is usually more efficient than procedure B.* Data features that significantly affect the relative performance of automatic integration procedures include the existence of sudden peaks in the graph of f, the existence of places where f's graph has no tangent line, whether *tol* is lenient (e.g., 10^{-3}) or stringent (e.g., 10^{-10}), and so on.

The only source of compelling evidence for such claims lies in careful measurements of program performance. Several difficult questions that arise when interpreting performance data are discussed in Section 6.3. Moreover, we will see that extensive measurements should be taken, because a small sample can easily be misleading.

Section 6.4 is based on the approach taken by J. N. Lyness and J. J. Kaganove in "A Technique for Comparing Automatic Quadrature Routines" (*Computer Journal*, 1977, pp. 170–177). We will see how extensive performance measurements can be condensed to produce easy-to-use graphs that depict the relative efficiency of automatic integration procedures for solving a family of integration problems.

While the chapter can be studied in isolation, a broader understanding of numerical software is needed to place this material in proper perspective. The reader who has skipped Chapters 2–5 can gather an appreciation of the issues we treat from "Comments on the Nature of Automatic Quadrature Routines," also by Lyness and Kaganove (*ACM Transactions on Mathematical Software*, March 1976, pp. 68–81). This paper is recommended reading for any student of numerical software, regardless of background.

6.1 SIMPSON'S RULE

This section discusses the properties of Simpson's rule that we will use in the construction of automatic integration procedures. Although Simpson's rule is not a particularly good choice for a basic integration formula (certain other formulas lead to more efficient automatic integration procedures), it has the advantage of simplicity. The general points made in this chapter are independent of which basic formula is used.

Simpson's rule is a traditional method for numerical integration, that is, for computing an approximation to $\int_c^d f(x)\,dx$ based on sample values of f. The intuitive idea is that while you can approximate the graph of a function if you know the value of the function at two points, you can approximate the graph somewhat better if you know the function's value at a third point midway between the first two. The interval between the two end points is referred to as a "panel." The midpoint can also be thought of as dividing the interval into two panels, and the rule can be applied recursively to obtain still more accurate results by evaluating the function at the midpoints of those panels.

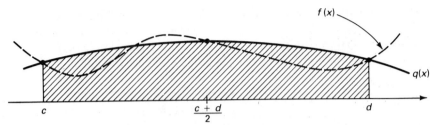

S_1 is the area of the shaded region

Figure 6.1

Simpson's rule with one panel is

$$S_1[f,c,d] = \frac{d-c}{6}\left[f(c) + 4f\left(\frac{c+d}{2}\right) + f(d)\right]$$

S_1 equals $\int_c^d q(x)\,dx$, where $q(x)$ is the parabola (that is, the polynomial of degree 2) that agrees with f at $c, (c+d)/2$ and d (Exercise 1). See Figure 6.1.

Simpson's rule with two panels is

$$S_2[f,c,d] = S_1\left[f,c,\frac{c+d}{2}\right] + S_1\left[f,\frac{c+d}{2},d\right]$$

$$= \frac{d-c}{12}\left[f(c) + 4f\left(\frac{3c+d}{4}\right) + 2f\left(\frac{c+d}{2}\right)\right.$$

$$\left. + 4f\left(\frac{c+3d}{4}\right) + f(d)\right]$$

See Figure 6.2.

A classical theorem of numerical analysis states, in effect, that if f is "sufficiently smooth" and if $d - c$ is "sufficiently small," then

$$\int_c^d f(x)\,dx - S_1[f,c,d] \approx 16\left(\int_c^d f(x)\,dx - S_2[f,c,d]\right)$$

That is, for short intervals, S_2 is about 16 times more accurate than is S_1. Thus S_1, S_2, and $I = \int_c^d f(x)\,dx$ should be related roughly as in Figure 6.3. This suggests that we "extrapolate" from S_1 and S_2 to the value

$$T = S_2 + \frac{S_2 - S_1}{15}$$

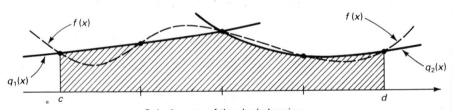

S_2 is the area of the shaded region

Figure 6.2

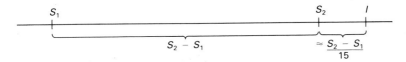

Figure 6.3

that may be a better approximation to the integral than is S_2. Rewriting the approximation as

$$I - S_2 \approx \frac{S_2 - S_1}{15}$$

provides a formula to estimate the error in S_2 using only S_1 and S_2.

Example 1
The following program applies Simpson's rule to integrate $f(x) = 0.01/(0.0001 + x^2)$ over the interval $[0,h]$, where h is repeatedly halved. See Figure 6.4.

Values of f are made available in the program by writing f as a "statement function." Elementary calculus shows that $\int_0^h f(x)\,dx = arctan(100h)$, which allows us to compare the computed integral with the correct value.

```
*THIS PROGRAM ILLUSTRATES THE CONVERGENCE OF S2 AND T TO
*THE INTEGRAL, AS H APPROACHES ZERO. IN ADDITION, IT
*ILLUSTRATES THE USEFULNESS OF |S2-S1|/15 AS AN ERROR
*ESTIMATOR.

      IMPLICIT DOUBLE PRECISION (A-H,O-Z)

      F(X) = 0.01D0 / (0.0001D0 + X**2)

      PRINT *,' H    ERROR IN S2    ERROR IN T    ESTIMATE'
      PRINT *
      DO 10 I = 1, 9
          INVERS = 2**I
          H = 1.0D0 / INVERS
          ANSWER = ATAN(100*H)
          S1 = (H / 6)*(F(0.0D0) + 4*F(H / 2) + F(H))
          S2 = (H / 12)*(F(0.0D0) + 4*F(H / 4) + 2*F(H / 2) +
     *         4*F(3*H / 4) + F(H))
          ESTIM = (S2 - S1) / 15
          T = S2 + ESTIM
          PRINT 1, INVERS, ABS(S2 - ANSWER), ABS(T - ANSWER),
     *         ABS(ESTIM)
1         FORMAT('1 /', I3, 3F14.8)
10    CONTINUE
      END
```

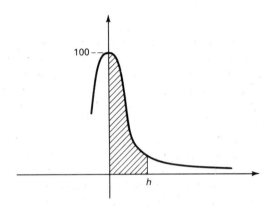

Figure 6.4

Here is the output (for a particular machine):

H	ERROR IN S2	ERROR IN T	ESTIMATE
1 / 2	2.74867773	2.47598171	.27269581
1 / 4	.81398571	.68501842	.12896720
1 / 8	.04323518	.00812411	.05135925
1 / 16	.08548844	.09402502	.00853663
1 / 32	.01786208	.01328564	.00457640
1 / 64	.00038695	.00082314	.00121008
1 / 128	.00000787	.00002545	.00001759
1 / 256	.00000232	.00000009	.00000241
1 / 512	.00000012	.00000001	.00000013

The important points to notice are these:

1. Both S_2 and T converge to $\int_0^h f(x)\,dx$ as h approaches 0, with T converging somewhat faster. Note, however, that for several values of h, S_2 is more accurate than is T.
2. For small h, $(S_2 - S_1)/15$ gives a good approximation to the error in S_2. For instance at $h = \frac{1}{512}$ we have $1.3 \times 10^{-7} \approx 1.2 \times 10^{-7}$.
3. In this example, T is more accurate when $h = \frac{1}{256}$ than is S_2 when $h = \frac{1}{512}$.

Example 2

Suppose that the statements in the program of Example 1 that define f and its integral are replaced by

```
F(X) = 0.01D0 / (0.0001D0 + (X - 1.0D0)**2)
ANSWER = ATAN(100*(H - 1.0D0)) - ATAN(-100.0D0)
```

Sec. 6.1 Simpson's Rule

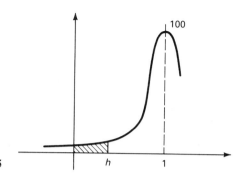

Figure 6.5

See Figure 6.5. The output becomes

H	ERROR IN S2	ERROR IN T	ESTIMATE
1 / 2	.00000834	.00000273	.00000561
1 / 4	.00000006	.00000000	.00000006
1 / 8	.00000000	.00000000	.00000000
	etc.		

The important points to notice are these:

1. Again, both S_2 and T converge to $\int_0^h f(x)\,dx$, with T converging somewhat faster. However, accurate approximations are obtained with larger values of h than in Example 1. The reason is that f is being integrated over a region where it is relatively flat, whereas Example 1 deals with a sharply curving section of a function.
2. Again, $(S_2 - S_1)/15$ gives a useful approximation to the error in S_2.
3. Again, using T instead of S_2 allows one to terminate the interval bisecting process sooner yet achieve the same accuracy.

PROGRAMMING ASSIGNMENT
(Optional)

Let $Q_1[f, c, d]$ be defined by

$$Q_1[f, c, d] = \frac{d - c}{90}\left[7f(c) + 32f\left(\frac{3c + d}{4}\right) + 12f\left(\frac{c + d}{2}\right)\right.$$
$$\left. + 32f\left(\frac{c + 3d}{4}\right) + 7f(d)\right]$$

Q_1 equals $\int_c^d q(x)\,dx$, where $q(x)$ is the polynomial of degree 4 that agrees with f at the following five values of x: c, $(3c + d)/4$, $(c + d)/2$, $(c + 3d)/4$, and d. Also, let $Q_2[f, c, d]$ be defined as

$$Q_2[f, c, d] = Q_1\left[f, c, \frac{c + d}{2}\right] + Q_1\left[f, \frac{c + d}{2}, d\right]$$

A classical theorem of numerical analysis states, in effect, that if f is "sufficiently smooth" and if $d - c$ is "sufficiently small," then

$$\int_c^d f(x)\,dx - Q_1[f,c,d] \approx 64\left(\int_c^d f(x)\,dx - Q_2[f,c,d]\right)$$

Mimic the computations discussed in this section but with S_1 and S_2 replaced by Q_1 and Q_2.

EXERCISES

1. Let $q(x) = Ax^2 + Bx + C$. Show that

$$\int_c^d q(x)\,dx = \frac{A(d^3 - c^3)}{3} + \frac{B(d^2 - c^2)}{2} + C(d - c)$$

$$= \frac{d - c}{6}\left[q(c) + 4q\left(\frac{c + d}{2}\right) + q(d)\right]$$

Derive the formula for S_2.

2. Show that the error estimate $(S_2 - S_1)/15$ simplifies to

$$\frac{d - c}{180}\left[-f(c) + 4f\left(\frac{3c + d}{4}\right) - 6f\left(\frac{c + d}{2}\right) + 4f\left(\frac{c + 3d}{4}\right) - f(d)\right]$$

3. Find a function $f(x)$ for which $S_1 = S_2 \neq I$. [This shows that the error estimate $(S_2 - S_1)/15$ is not completely reliable.]

6.2 A SIMPLE PROCEDURE FOR AUTOMATIC INTEGRATION

A "divide-and-conquer" approach, based on observations made in the previous section, leads to an automatic integration procedure

$$\text{area}(f, a, b, tol)$$

that attempts to compute an approximation to $\int_a^b f(x)\,dx$. The interval of integration, $[a, b]$ is automatically divided into a set of subintervals $[a, x_1], [x_1, x_2], \ldots, [x_k, b]$, using small intervals where f is difficult to integrate and large intervals where f is easy to integrate.

The basic idea of the algorithm is as follows. If we estimate that the error in $S_2 = S_2[f, a, b]$ does not exceed tol, then we will set $\text{area} \leftarrow S_2$. Otherwise,

Sec. 6.2 A Simple Procedure for Automatic Integration

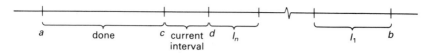

Figure 6.6

we will divide the interval $[a, b]$ in half, integrate the two pieces separately, each with tolerance $tol/2$, and add the results.

Algorithm 6.1 (Recursive Version)

$\Theta \leftarrow 15$ ..(Just one possible setting.)

Compute $S_1 = S_1[f, a, b]$ and $S_2 = S_2[f, a, b]$.

if $|S_2 - S_1|/\Theta \leq tol$

..Accept S_2 as an approximation to $\int_a^b f(x)\,dx$.

..[Another possibility is $S_2 + (S_2 - S_1)/15$.

..Perhaps a larger value of Θ can then be used.]

area $\leftarrow S_2$

else

$$\text{area} \leftarrow \text{area}\left(f, a, \frac{a+b}{2}, \frac{tol}{2}\right) + \text{area}\left(f, \frac{a+b}{2}, b, \frac{tol}{2}\right)$$

Algorithm 6.1 can be implemented fairly easily, even in a programming language like FORTRAN that does not allow a procedure to invoke itself. At any time during execution of the procedure, a "current interval" $[c, d]$ will be under consideration, a seemingly accurate approximation to $\int_a^c f(x)\,dx$ will have been accepted, and there will exist n "pending" intervals to the right of $[c, d]$ that await further processing. See Figure 6.6. The user-supplied tolerance tol is distributed among the intervals in proportion to their length, so the acceptable error allotted to a subinterval $[c, d]$ is $tol \times [(d - c)/(b - a)]$.

Algorithm 6.1 (Iterative Version)

$\Theta \leftarrow 15$ ..(Just one possible setting.)

current interval $\leftarrow [a, b]$

area $\leftarrow 0$

$n \leftarrow 0$

repeat

Let $[c, d]$ denote the current interval.

Compute $S_1 = S_1[f, c, d]$ and $S_2 = S_2[f, c, d]$.

if $|S_2 - S_1| / \Theta \le tol \times \dfrac{d-c}{b-a}$

..Accept S_2 or T as an approximation to $\int_c^d f(x)\, dx$ ...

area ← area + S_2

.. and work on the next pending interval.

if $n = 0$

    done

else

    current interval ← I_n

    $n \leftarrow n - 1$

else

..Save the right half...

$n \leftarrow n + 1$

$I_n \leftarrow \left[\dfrac{c+d}{2}, d \right]$

..and work on the left half.

current interval ← $\left[c, \dfrac{c+d}{2} \right]$

Implementation Notes

1. It may be desirable to implement Algorithm 6.2 using *reverse communication*. With this mechanism, which is closely related to the notion of a "coroutine," the integration procedure returns to the calling program whenever it needs a value of f. Procedures implemented this way are somewhat cumbersome to write. (For instance, FORTRAN's rule against branching into the range of a *DO* statement may make it necessary to write loops with explicit *GOTO*s.) However, they are often substantially easier to use than are procedures invoked with the name of an external integrand function as an argument. For instance, see Exercise 2 or compare the program in Section 5.3 with the one in Section 6.4. Reverse communication may be appropriate for a procedure written by an expert programmer and intended for use by unsophisticated programmers. It is wise to expend a few minutes of the expert's time to spare naive users of having to cope with the "strange" notions of *COMMON* blocks and *EXTERNAL* functions.

A FORTRAN integrator implemented with reverse communication might be invoked as follows to compute

$$\int_0^1 \dfrac{0.01}{0.0001 + x^2}\, dx$$

Sec. 6.2 A Simple Procedure for Automatic Integration

```
      IFLAG = 0
10    FOUND = AREA (0.0, 1.0, X, FX, IFLAG, TOL)
      IF (IFLAG .EQ. 1) THEN
         FX = .01 / (.0001 + X**2)
         GO TO 10
      END IF
*     (..AT THIS POINT IFLAG = 2)
```

Successful use of reverse communication depends on the values of the integrator's local variables being preserved between a return to the calling program and subsequent reentry. The FORTRAN 77 *SAVE* statement is useful in this regard. Particular care may be required with the variable *AREA*.

Although this discussion is sketchy, readers have preferred to use reverse communication in Programming Assignment 1 without a more extensive hint. You, too, will be able to implement Algorithm 6.1 this way if you try.

2. The operation

$$I_n \leftarrow \left[\frac{c+d}{2}, d\right]$$

means storing certain values (end points, values of f, etc.). Suppose that there are k such values associated with each interval; call them $val_1, val_2, \ldots, val_k$. Using arrays $save_1, save_2, \ldots, save_k$, the assignment

$$I_n \leftarrow \left[\frac{c+d}{2}, d\right]$$

is accomplished by

$$save_1(n) \leftarrow val_1; \ldots; save_k(n) \leftarrow val_k$$

The assignment

$$current\ interval \leftarrow I_n$$

is realized by

$$val_1 \leftarrow save_1(n); \ldots; val_k \leftarrow save_k(n)$$

The results of Exercise 1 suggest that it is reasonable to allocate p words of storage for each of these arrays $save_i()$, where p is the number of bits in a floating-point fraction. If the procedure reaches the point that there are p pending intervals, it can avoid stack overflow by automatically accepting S_2.

3. Precautions should be taken so that the integration procedure cannot consume a disastrous amount of time. For instance, if 5000 values of f have been computed, then the integration procedure could stop with an error message. Alternatively, it might switch to a strategy that takes only a small number of further samples, does as best it can with that information, and returns a warning flag. In particular, the iterative version of Algorithm 6.1 might just accept S_2 for each of the n pending intervals. For initial program

checkout we suggest that the maximum number of function invocations be set lower, say to 500.

Typographical Mistakes. Experience with students' integration programs led us to formulate the following hypothesis:

Hypothesis 6.1
A mistake in an implementation of Algorithm 6.1 is likely to cause the program to invoke f the specified maximum number of times.

We conducted a mutation experiment on (1) a FORTRAN implementation of Algorithm 6.1, (2) the integral

$$\int_0^1 \frac{0.01}{0.0001 + (x - 0.6)^2} \, dx$$

and tolerance $tol = 10^{-3}$, and (3) the acceptance criteria that (i) the computed integral lie within 10^{-3} of the exact value 3.09993... and that (ii) the integrator invoke f at most 500 times. [The original integrator (1) used 89 function values.] 83 of the 1767 mutants satisfied both criteria. Of the 1684 mutants that failed, 800 requested too many function values, 471 referenced a variable that had not been assigned a value, 234 produced an approximate integral that was not sufficiently accurate, 147 used an illegal array subscript, and 32 committed other transgressions. These results underscore the importance of writing your integration procedure so that it cannot exceed a fixed number of function invocations.

The experiment turned up another point that is worth mentioning. Most of the 83 mutants that survived the experiment are extremely hard to detect with performance measurements. For instance, lowering the permissible number of function evaluations from 500 to 100 resulted in the detection of only four additional typographical mistakes.

In particular, certain parts of the integration program proved to be particularly susceptible to typographical mistakes that are difficult to detect; the corresponding parts of your program deserve extra scrutiny. Mistakes in portions of the program that are executed only once can be masked by the other computations. (Mistakes in unexecuted statements, e.g., those handling stack overflow, will obviously go undetected.) Moreover, mistakes in the convergence test or the extrapolation formula can be quite hard to detect. For instance, the typographical change from

```
IF ( ABS(S2-S1) .LE. 15*TOL*(D-C)/(B-A) )
```
to
```
IF ( ABS(S2-S1) .LE. 15*TOL*(D-C)/(B-S1) )
```
decreased the number of function evaluations from 89 to 81 and raised the

error from $10^{-3} \times 0.135\ldots$ to $10^{-3} \times 0.137\ldots$. Changing

AREA = AREA + S2 + (S2 - S1) / 15

to

AREA = AREA + S2 - (S2 - S1) / 15

increased the error from $10^{-3} \times 0.135\ldots$ to $10^{-3} \times 0.544\ldots$ while preserving the cost.

PROGRAMMING ASSIGNMENTS

1. Implement Algorithm 6.1. Values of f should be saved until they are no longer useful. It should be possible to prove that your program will never generate an illegal subscript and will never consume an inordinate amount of time.

2. (Optional) Implement Algorithm 6.1 but use the basic integration rule suggested by the Programming Assignment of Section 6.1 instead of Simpson's rule.

3. (Optional) Design and implement an automatic integration procedure that works as follows. The formulas for approximating an integral and for estimating the error in that approximation will be the same as in Algorithm 6.1. At each stage, the interval with the largest estimated error will be divided in half. When the sum of the absolute values of the error estimates for all pending intervals falls below *tol*, the estimated integrals over all pending intervals will be added together to give the result.

   This approach requires a data structure for the pending intervals in which you can find the interval with largest estimated error, delete that interval, and insert the two halves. An exceptionally fancy solution to this assignment will perform each of these operations in time $O(\log n)$, where n is the number of pending intervals. (A FORTRAN program that attains these goals using a "heap of pointers" has been given in "Local Versus Global Strategies for Adaptive Quadrature" by Michael Malcolm and R. Bruce Simpson, *ACM Transactions on Mathematical Software*, June 1975, pp. 129–146.) However, for the purposes of this assignment and Programming Assignment 3 of Section 6.4, a simple data structure is adequate.

EXERCISES

1. (a) For the iterative version of Algorithm 6.1, show that $d - c \leq 2^{-n}(b - a)$, where $[c, d]$ is the current interval, n is the number of pending intervals, and $[a, b]$ is the initial interval of integration. [*Hint*: Prove a stronger result by induction on the number of times that the *repeat* loop has been executed.]

   (b) Let p be the number of bits in a floating-point fraction. Show that if there are p pending intervals, then the length of the current interval is less than one ulp of $b - a$.

2. Suppose that you are implementing an interactive language for solving mathematical problems. In particular, suppose that your program is to handle a command like

   INTEGRATE F(X) = 0.01 / (0.0001 + X**2) FROM 0.0 TO 1.0

with the help of a packaged integration procedure. Outline the general steps that your program must take if the integrator uses reverse communication. What if the integrator expects the name of an external function f as one of its arguments, as in procedure area(f, a, b)?

6.3 MEASURING THE PERFORMANCE OF INTEGRATION PROCEDURES

In this Section we will discuss performance measurements of automatic integration procedures. In particular, the following points will be observed:

1. Any automatic integration procedure will occasionally fail.
2. No specification exists for automatic integration procedures.
3. We must decide how to treat measured program failures.
4. Performance can be quite sensitive to small changes in the program input.

6.3.1 Any Automatic Integration Procedure Will Occasionally Fail

It is impossible for a numerical method to infallibly compute $\int_a^b f(x)\,dx$ using only a finite number of function samples and no other information about f. See Figure 6.7.

Another way to view this limitation is that any given numerical integration procedure can be "fooled" as follows. Run the procedure with $a = 0$ and $b = 1$, returning 0 every time it asks for a function sample and recording the arguments $x_1, x_2, \ldots, x_k$ at which it requests these values. After the procedure returns a computed integral, presumably 0, you can say, "Gotcha!" and report the procedure's dismal failure at integrating the polynomial

$$f(x) = 7104368(x - x_1)^2(x - x_2)^2 \cdots (x - x_k)^2$$

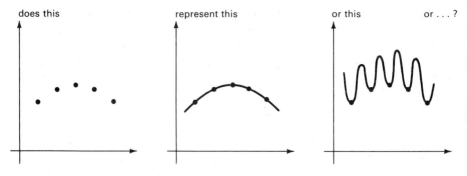

Figure 6.7

Sec. 6.3 Measuring the Performance of Integration Procedures

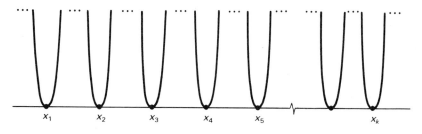

Figure 6.8

See Figure 6.8. Notice that you need to be able to call the integration procedure to construct this function but you do not need to know how it works.

6.3.2 The Lack of a Specification

Developers of numerical software have been unable to find specifications, in the sense of Section 1.1, for automatic integration procedures. The precise statements about input/output behavior that *can* be made [for example, that $\int_a^b f(x)\,dx$ is computed exactly if f is a polynomial of degree 3 and if exact arithmetic is used] are woefully inadequate. In other words, automatic integration procedures work far better in practice than we know how to state in a precise way.

On the other hand, it is possible to make *useful* (though limited) assertions about program performance, and the program can be run to corroborate those claims. For example, computing $\int_0^1 f(x)\,dx$, where

$$f(x) = \begin{cases} x^{-\frac{1}{2}} & \text{if } x > 0 \\ 0 & \text{if } x \leq 0 \end{cases}$$

(the exact integral is 2) will probably exercise the "stack overflow" mechanism of an implementation of Algorithm 6.1 and thereby check the assertion that the program's array subscripts stay in bounds. Similarly, one might run an integration procedure on a function f whose values are random to check that the procedure will not request an exorbitant number of function values.

6.3.3 A Decision About Program Failures

Developers of numerical software have learned the hard way that performance measurements of integration procedures can be difficult to interpret.

For an introduction to the difficult questions that arise, consider the following integrals. (The given values were computed using the approximation 3.14159 for π.)

$$\int_0^1 \frac{2}{2 + \sin(10\pi x)} \, dx \approx 1.15470066904 \tag{1}$$

$$\int_{0.01}^1 \frac{[\sin(50\pi x)]^2}{(\pi x)^2} \, dx \approx 0.11213956963 \tag{2}$$

$$\int_{0.1}^1 \frac{\sin(100\pi x)}{\pi x} \, dx \approx 0.9098645256 \times 10^{-2} \tag{3}$$

The following performance results involving these integrals are taken from "Comparison of Numerical Quadrature Formulas" by D. K. Kahaner (in *Mathematical Software*, edited by John R. Rice, Academic Press, New York, 1971, pp. 229–259).

Problem	Integral	tol	SIMPSN Error	SIMPSN Cost	QNC7 Error	QNC7 Cost
1	1	10^{-3}	0.72×10^{-4}	163	0.89×10^{-9}	97
2	2	10^{-3}	0.83×10^{-3}	151	0.11×10^{-2}	169
3	2	10^{-6}	0.82×10^{-7}	2275	0.11×10^{-2}	385
4	3	10^{-6}	0.91×10^{-2}	19	0.99×10^{-11}	1525

On problem 1, the program *QNC7* clearly outperforms *SIMPSN* (which is essentially an implementation of Algorithm 6.1) since *SIMPSN* uses about 65% more function values. But what weight should we give to the fact that *QNC7* produced a solution that was far more accurate than the solution produced by *SIMPSN*? One school of thought holds that an integration procedure should be rewarded for attaining higher accuracy than is requested, while another disagrees.

On problem 2, *SIMPSN* performs better than *QNC7*, but it is a matter of opinion whether *SIMPSN* performs a little better or a lot better. Specifically, how much weight should we give to the fact that the error in the solution computed by *QNC7* exceeded *tol*? Is the fact that *QNC7* fails by a mere 10% any more excusable than a more extreme failure?

On problem 3, *SIMPSN* is successful, while *QNC7* fails utterly. However, the roles are reversed on problem 4. Is an utter failure after 385 function evaluations any worse than an utter failure that costs only 19 evaluations?

There are no easy answers to these questions.

Here we will adopt the viewpoint that an integration procedure's *efficiency* (that is, the relationship between actual error and cost) is an issue different from its *reliability* (that is, the relationship between actual error and specified tolerance), and our effort will be focused on measuring efficiency. This approach is at least partially justified by the observation that making an integration procedure more efficient is often difficult (or impossible), whereas making it more reliable is usually trivial. For instance, making Algorithm 6.1 more efficient might involve, say, changing from approximation with quadratics to the use of higher-degree polynomials. (However, such a change may make the procedure less efficient.) On the other hand, Algorithm 6.1 can be made more reliable by changing the assignment $\Theta \leftarrow 15$ to, say, $\Theta \leftarrow 10$. (See Exercise 1.) The same effect is obtained by having the integrator internally reset *tol*. Moreover, the data-summarizing techniques developed in the next section can be applied directly to reliability measurements.

A number of systematic comparisons of integration procedures have been conducted, and the amount of performance data generated is often staggering. For instance, in the study mentioned above, Kahaner measured 11 automatic integration routines on 21 combinations of (f, a, b), with three values of *tol*. For each of the $11 \times 21 \times 3 = 693$ runs Kahaner recorded the error, the number of function evaluations, and the elapsed time. Merely listing these measurements takes 11 pages, but as is suggested by the results in the table above, the evidence does not allow one to claim that any one procedure is "best," or even uniformly good. Each integration procedure looked relatively efficient on some problems and inefficient on others.

In some cases, graphs provide a way of presenting large accumulations of performance data in a form that can be assimilated quickly. For instance, consider the results in Figure 6.9 obtained by applying an implementation of Algorithm 6.1 with various values of *tol* to the integral

$$\int_0^1 \frac{0.01}{0.0001 + (x - \alpha)^2} \, dx$$

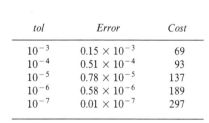

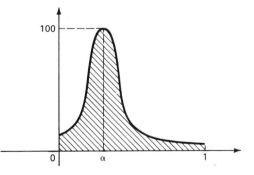

Figure 6.9

with $\alpha = 0.2811535$. Plotting each of the (*error*, *cost*) points and connecting them with line segments produces Figure 6.10. Notice that Figure 6.10 depicts cost versus actual error; the value of *tol* that determined a point is not indicated. This fact reflects our decision to consider efficiency, not reliability.

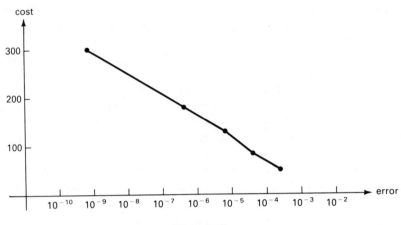

Figure 6.10

6.3.4 The Sensitivity of Performance

Setting α to 0.55938008 (Figure 6.11) and repeating the above measurement process, one obtains the following performance data:

tol	Error	Cost
10^{-3}	7.5×10^{-3}	57
10^{-4}	82.8×10^{-4}	81
10^{-5}	832.3×10^{-5}	109
10^{-6}	2.2×10^{-6}	193
10^{-7}	1.0×10^{-7}	301

Graphed with the earlier results, these data appear as shown in Figure 6.12. The point to note is that on two quite similar integration problems, the procedure has very different performance characteristics. The lesson to be

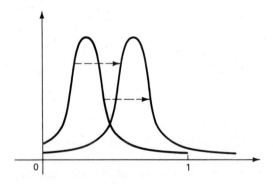

Figure 6.11

Sec. 6.4 Sturdier Cost-Versus-Error Graphs   155

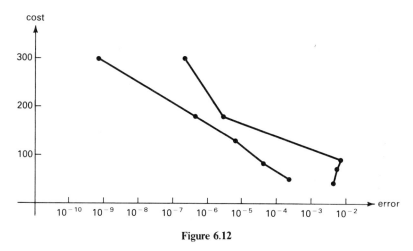

Figure 6.12

learned is that it is risky to use performance measurements obtained on one sample integral to predict program performance on similar integrals.

PROGRAMMING ASSIGNMENT

Measure the reliability of your integration procedure with several values of the threshold Θ. Does the extrapolated approximation $S_2 + (S_2 - S_1)/15$ seem more efficient than S_2?

EXERCISE

1. Explain why altering the assigned value of Θ in Algorithm 6.1 changes the procedure's reliability but not its efficiency.

6.4 STURDIER COST-VERSUS-ERROR GRAPHS

This section considers cost-versus-error graphs that look like the ones used in Section 6.3. However, because they are obtained by averaging measurements from a large number of runs, the graphs discussed here are relatively free of the sensitivity to minor changes in f, a, and b that plagues the earlier approach. In addition to increasing the likelihood that the measurements will accurately describe the procedure's behavior for a meaningful class of problems, this approach has the advantage of making explicit the statistical nature of our claims about program performance.

The two approaches to constructing cost-versus-error graphs have much in common. In both cases, we begin by picking a few values of *tol*; call them

$tol_1, tol_2, \ldots, tol_m$. Corresponding to each of these m values there is a point on the graph that is determined by measuring the performance of the integration procedure when invoked with that value of *tol*. (The graph does not indicate the value of *tol* that determines the point.) The graph is obtained by connecting the m points with line segments.

In Section 6.3, the coordinates of the point corresponding to tol_i were merely the error and cost found by solving a fixed integration problem $\int_a^b f(x)\,dx$. Intuitively, a point on the graph gives the cost of computing an approximate integral with a specified error.

In the approach described in this section, we will set the integration procedure's tolerance to tol_i and record the error and cost for each of n members of a family of extremely similar integration problems. Let us denote by err_j and $cost_j$ the error and cost for the jth of these n problems. These measurements will be summarized as a point $(ERROR_i, COST_i)$ on our cost-versus-error curve.

A Family of Integration Problems. Our discussion will center on integration problems of the form $\int_0^1 f_\alpha(x)\,dx$, where

$$f_\alpha(x) \frac{0.01}{0.0001 + (x - \alpha)^2}$$

See Figure 6.13. However, the underlying principles can be applied to any one-parameter family of computing problems. (This is called a "one-parameter family" of problems because each value of the parameter α determines a different integral.) These specific functions $f_\alpha(x)$ are particularly useful for measuring the performance of integration procedures because their graphs have both flat and curvy sections and because their integrals have the closed-form expression

$$\int_a^b f_\alpha(x)\,dx = arctan[100(b - \alpha)] - arctan[100(a - \alpha)]$$

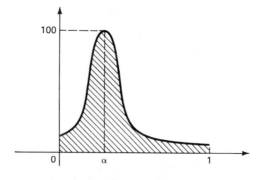

Figure 6.13

Sec. 6.4 Sturdier Cost-Versus-Error Graphs

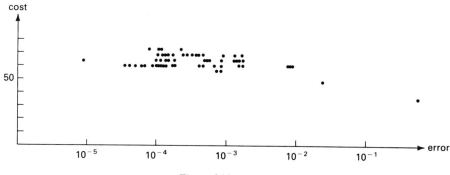

Figure 6.14

so the correct value can be computed easily. Many numerical experiments have been carried out whose specific purpose is to show that the performance of an integration procedure on these functions is indicative of its performance on smooth functions with isolated peaks. [However, performance on these problems need not be a good measure of the procedure's ability to integrate a function whose graph is not smooth, or wildly oscillating functions like $sin(100x)$.]

Summarizing the Measurements for a Fixed Tolerance. A scatter plot of $(err_j, cost_j)$ pairs obtained with $tol = 10^{-3}$ is given in Figure 6.14. The measurements were taken by applying an implementation of Algorithm 6.1 to $n = 100$ randomly chosen members of the above problem family. (About 15% of the integrations failed to obtain the requested accuracy. This failure rate is typical for reasonably cautious integration procedures on fairly hard problems.) The spread of errors, roughly 10^{-5} to 1.0, underscores the danger in basing general claims about program performance on a few isolated trials.

Cost typically varies much less erratically with small changes in f than does the error. For that reason we will be content to consider only the average cost and thus will set

$$COST_i = \frac{\sum_{j=1}^{n} cost_j}{n}$$

$ERROR_i$ will be determined as follows. Fix a confidence level s, for example, $s = 80$, $s = 90$, or $s = 95$. $ERROR_i$ is defined as the smallest number that lies to the right of (i.e., is at least as large as) $s\%$ of the n measured errors err_j. For example, if $n = 10$ and $s = 90$, then $ERROR_i$ is the second largest err_j. See Figure 6.15.

The point $(ERROR_i, COST_i)$ has the following intuitive interpretation. To be $s\%$ certain of computing a value within $ERROR_i$ of the answer for some member of this family of integration problems, you can expect to pay about $COST_i$ function values. In particular, one can attain the $s\%$ level of confidence

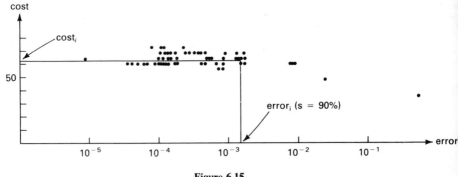

Figure 6.15

on the measured problems by setting the procedure's tolerance to tol_i, but this value of tol is not depicted in the cost-versus-error graph.

This intuitive interpretation of points on our error-versus-cost graph encourages us to adopt a new attitude toward tol. In particular, it suggests that beside determining a maximum acceptable error, the sophisticated user should determine the level of risk that he or she is willing to take. The user can then employ tol as a kind of "tuning knob" that may well be set at a value different from the maximum acceptable error.

Computing the Cost-Versus-Error Graph. The following program generates 18 points on the cost-versus-error graphs for $s = 80$, 90, and 95. The points are those determined by the tolerances $10^{-tolexp}$, with $tolexp = 1.5, 2.0, 2.5, \ldots, 9.5, 10.0$. The program assumes that the integration procedure is implemented using "reverse communication" as described in implementation note 1 of Section 6.2. Notice how this simplifies the jobs of changing α and recording the number of function evaluations.

```
      IMPLICIT DOUBLE PRECISION (A-H, O-Z)
      DOUBLE PRECISION ERR(100)
      INTEGER COST(100)
*  N IS THE NUMBER OF INTEGRATION PROBLEMS SOLVED WITH A
*  FIXED TOLERANCE.
*  M IS THE NUMBER OF COST-VERSUS-ERROR POINTS TO BE
*  DETERMINED.
      N = 100
      M = 18
      LEV80 = .80*N
      LEV90 = .90*N
      LEV95 = .95*N
      PRINT *, 'TOL   S = 80   S = 90   S = 95   COST'
      PRINT *
```

Sec. 6.4   Sturdier Cost-Versus-Error Graphs

```
              DO 30 I = 1, M
                    TOLEXP = (I + 2) / 2.0D0
                    TOL = 10.0D0**( - TOLEXP)
                    CALL MEASUR (TOL, ERR, COST, N)
                    CALL LARGST (ERR, N - LEV80 + 1, N)
                    AVCOST = 0.0
                    DO 10 J = 1, N
10                        AVCOST = AVCOST + COST(J)
                    AVCOST = AVCOST / N
                    PRINT 20, TOL, ERR(LEV80), ERR(LEV90),
     *              ERR(LEV95), AVCOST
20                  FORMAT (4E9.3, F8.2)
30            CONTINUE
              END


*  MEASUR - GATHER MEASUREMENTS OF AN INTEGRATION ROUTINE.
      SUBROUTINE MEASUR (TOL, ERR, COST, N)
      IMPLICIT DOUBLE PRECISION (A - H, O - Z)
      DOUBLE PRECISION ERR(*)
      INTEGER COST(*)
      G(Z) = ATAN(100.0D0*(Z - ALPHA))

              DO 20 I = 1, N
*                   GENERATE A RANDOM NUMBER BETWEEN 0 AND 1.
                    ALPHA = RAN()
                    COST(I) = 0
                    IFLAG = 0
10                  FOUND = AREA(0.0D0,1.0D0,X,FX,IFLAG,TOL)
                    IF (IFLAG .EQ. 1) THEN
                          FX = 0.01D0 / (0.0001D0 +
     *                    (X - ALPHA)**2)
                          COST(I) = COST(I) + 1
                          GO TO 10
                    ENDIF
                    ANSWER = G(1.0D0) - G(0.0D0)
                    ERR(I) = ABS(ANSWER - FOUND)
20            CONTINUE
              RETURN
              END


*  LARGST - MOVE THE LARGEST K ELEMENTS TO THE REAR OF
*  AN ARRAY.
```

```
SUBROUTINE LARGST (A, K, N)
DOUBLE PRECISION A(N), TEMP
DO 20 ITIMES = 1, K
        LIM = N - ITIMES + 1
        LOC = 1
        DO 10 I = 2, LIM
                IF (A(I) .GT. A(LOC)) LOC = I
10      CONTINUE
        TEMP = A(LIM)
        A(LIM) = A(LOC)
        A(LOC) = TEMP
20   CONTINUE
        RETURN
        END
```

Be warned that this program is expensive to run. See Exercise 1.

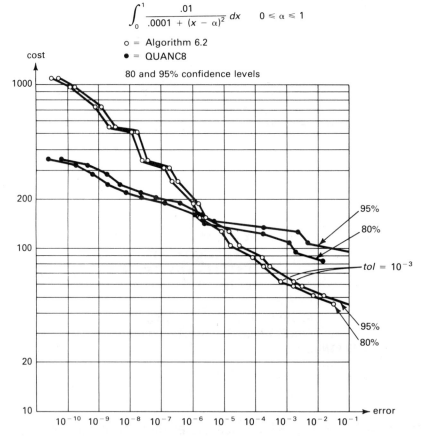

Figure 6.16

Sec. 6.4 Sturdier Cost-Versus-Error Graphs

Using the Cost-Versus-Error Graph. In Figure 6.16 we have graphed the pairs corresponding to $s = 80$ and $s = 95$ obtained by running the above program to measure the efficiency of an implementation of Algorithm 6.1. This determines two similar curves, with the curve corresponding to $s = 95$ naturally lying above the curve for $s = 80$.

The relative performance of two integration procedures on a given problem family can be determined by comparing the two corresponding cost-versus-error graphs. This comparison is greatly facilitated by the fact that, in practice, the results are essentially independent of s. For example, Figure 6.16 also depicts the graph for an integration procedure, called *QUANC8*, taken from the book *Computer Solution of Mathematical Problems* by Forsythe, Malcolm, and Moler (Prentice-Hall, Englewood Cliffs, N.J., 1977). (*QUANC8*

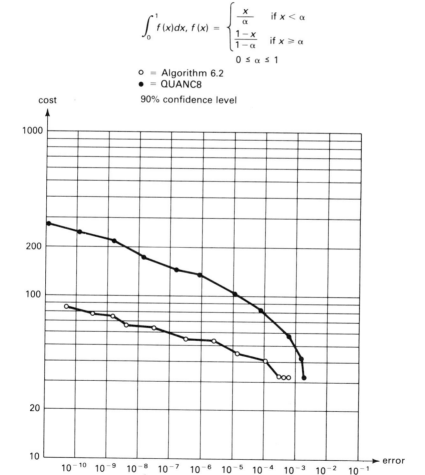

Figure 6.17

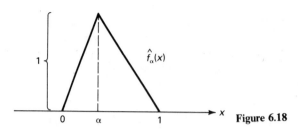

Figure 6.18

is similar to Algorithm 6.1 in structure. The main difference is that it uses a formula based on eighth-degree polynomials instead of Simpson's rule.) The following facts about the relative behavior of the two integration routines *on this problem family* are readily apparent. For low accuracy requirements, this implementation of Algorithm 6.1 is more economical than $QUANC8$. For higher accuracy, $QUANC8$ is better. For instance, to attain an error of 10^{-10} with 95% confidence costs about 1000 function evaluations with Algorithm 6.1 but only a third as much with $QUANC8$. The crossover point, where Algorithm 6.1 and $QUANC8$ are equally efficient, occurs for errors around $10^{-5.5}$. *Notice that these conclusions are valid for both $s = 80$ and $s = 95$.*

The results of such comparisons apply only to problems that closely resemble the members of the problem family used for the measurements. In general, for any two given integration procedures P_1 and P_2, there will exist a problem family on which P_1 outperforms P_2 (and vice versa). For instance, Figure 6.17 gives the results of measuring our implementation of Algorithm 6.1 and $QUANC8$ on the family $\int_0^1 \hat{f}_\alpha(x)\,dx$, with $\hat{f}_\alpha$ defined as in Figure 6.18. The results displayed in Figure 6.17 may have some wider validity for simple integrands f with a discontinuous derivative, but one should be cautious about making such generalizations.

PROGRAMMING ASSIGNMENTS

1. Adapt the techniques of this section to produce error-versus-tolerance (i.e., reliability) graphs. For reasons of economy you might perform $n = 10$ integrations for each of $m = 6$ values of *tol*, say $tol = 10^{-3}, 10^{-4}, \ldots, 10^{-8}$. Use the measurements to "tune" the integrator's criterion for accepting an approximate integral. [*Hint*: The implementation of Algorithm 6.1 used to produce Figures 6.16 and 6.17 accepts $S_2 + (S_2 - S_1)/15$ if $|S_2 - S_1| \le 30 \times [(d-c)/(b-a)] \times tol^{0.8}$.]

2. (Optional) The measurements summarized by Figure 6.16 support the following claim: *For integrating smooth functions f to stringent tolerances, automatic integration procedures using high-degree polynomial approximations are usually more efficient than those using low-degree polynomials. For lenient tolerances, low-degree polynomials may be more efficient.* Use performance measurements to further investigate this claim. Compare your implementation of Algorithm 6.1 with a

Sec. 6.4   Sturdier Cost-Versus-Error Graphs

procedure that uses an integration formula based on higher-degree polynomials (such as a solution to Programming Assignment 2 of Section 6.2).

3. (Optional) Use the techniques of this section to compare the performance of an implementation of Algorithm 6.1 with that of a solution to Programming Assignment 3 of Section 6.2.

4. (Optional) Repeat one of the immediately preceding programming assignments using the integrals $\int_0^1 h_\alpha(x)\,dx$, where

$$h_\alpha(x) = 100\,\text{sech}[\gamma(x - \alpha)] \quad \text{for } \gamma = 100\,ln(2 + \sqrt{3}\,).$$

The graph of h_α is very similar to that of f_α; indeed, γ is chosen so that

$$h_\alpha(\alpha \pm 0.01) = 50 = f_\alpha(\alpha \pm 0.01).$$

Letting

$$g(z) = (100/\gamma)\,arcsin(tanh(\gamma(z - \alpha))),$$

we have $\int_0^1 h_\alpha(x)\,dx = g(1) - g(0)$. Are the conclusions you reached using the functions f_α indicative of program behavior for this problem family?

EXERCISE

1. By examining Figure 6.16, estimate within 25% the number of function values that were needed to produce it.

RESEARCH PROJECT

Gather together a few of the best available automatic integration procedures. Find a modest-sized and complete set of (justifiable) claims of the following form: *For data (f, a, b, tol) with the features..., procedure A is usually the most efficient.* In other words, find a reasonably small set of problem "features" that allows one to successfully predict which of the procedures will be most efficient.

INDEX

A
algorithm, 2
 for automatic integration, 145
 for linear equations, 78–79, 81
 for minimization, 132
 for nonlinear equations, 111, 117, 124, 126–27
 for quadratic equations, 8
 for the sine function, 50
Anderson, N., 16, 101
argument purification, 68
argument reduction (for sine procedures), 47
automatic integration problem, 138

B
backward error analysis, 43
base, floating-point, 18
Basili, V., 126
bisection method, 20, 21, 111
Björck, A., 16, 101
blunder, 3
Brent, R., 105, 133
Brown, W., 27
Budd, T., 4
Bunch, J., 73
Bus, J., 125

C
Chandrasakeran, B., 4
Cody, W., 21, 28, 46, 49, 62, 67
condition number, 73
Cowell, W., 16, 27

D
Dahlquist, G., 16, 101
Davenport, S., 16
De Millo, R., 46
Dekker, T., 125
Delves, L., 16
derivative, relative, 60
Dongarra, J., 73, 86
DuCroz, J., 85

E
efficiency of an automatic integration procedure, 152
environmental parameters, 20
error, 3
Evans, D., 15
experiment, 2
 about linear equation solvers, 100, 103
exponent, floating-point, 18

F
Fast Givens method, 101
Fike, C., 49
floating-point standards, 28
Forsythe, G., 28, 73, 161
Fosdick, L., 16
Fox, P., 25
fraction, floating-point, 18

165

G

Gaussian elimination, 76–79, 84
  with complete pivoting, 108
  with partial pivoting, 78
Gauss-Jordan elimination, 81
Gentleman, W., 24, 101
George, A., 87
Gerhart, S., 46
golden section search, 130–32
Gram-Schmidt method, 101
guard digit (for subtraction), 35, 64, 70

H

Hall, A., 25
Hanson, R., 86
Hennel, M., 16
Hinds, A., 86
Horner's rule, 49
Householder's method, 101
Howden, W., 16
hypothesis, 2
  about automatic integration procedures, 148
  about linear equation solvers, 94, 103
  about nonlinear equation solvers, 118–19
  about quadratic equation solvers, 9, 10
  about sine procedures, 51

I

ill-conditioned linear equations, 75, 102, 105
implementation, 2

J

Jacobs, D., 16
Johnson, D., 83

K

Kaganove, J., 139
Kahaner, D., 152–53
Kernighan, B., 16
Kincaid, D., 86
Knuth, D., 5, 28, 30
Kreitzberg, C., 16
Krogh, F., 86

L

Larmouth, J., 16
Lawson, C., 86
linear equations, 72
linear least squares problem, 99
Linnainmaa, S., 105
LINPACK, 73, 83, 85–97
Lipton, R., 46
Liu, J., 87
Lyness, J., 139

M

machine epsilon, 19
macro processor, 27, 88
major interval, 130
Malcolm, M., 21, 28, 149, 161
Marovich, S., 24

maximum floating-point exponent, 18
Miller, W., 13, 27, 83, 100, 103–4
Mills, H., 126
minimization problem, 129
minimum floating-point exponent, 18
Minnihan, B., 83
minor interval, 130
mistake, 3
Moler, C., 28, 73, 161
most important fact about floating-point numbers, 19
Murphy's law of floating-point arithmetic, 17, 24, 46
mutant, 4
mutation experiment, 4
  about automatic integration procedures, 148
  about linear equation solvers, 94
  about nonlinear equation solvers, 118, 120
  about quadratic equation solvers, 10
  about sine procedures, 51

N

normal equations, 101
Nugent, S., 85
numerical oversight, 3

O

operation counts, for linear equation solvers, 79–80
overflow (floating-point), 19, 87
overflow (integer), 23, 53
oversight, 3

P

parameters, environmental, 20
Parlett, B., 73, 83
performance measurement, 2
Perlis, A., 46
Peters, G., 102
PFORT Verifier, 25
plane rotations, 101
Plauger, P., 16
PORT library, 24–27
portability oversight, 3
precision, floating-point, 18

Q

quadratic equation, 7

R

Radicci, S., 4
radix, floating-point, 18
random number generator, 5, 28–30
Reid, J., 16, 27, 28, 85
relative derivative, 60
reliability of an automatic integration procedure, 152
reverse communication, 146
Rice, J., 15, 16, 99, 152
root bracketing problem, 12, 109
root of a function, 109

Index

root refining problem, 110
Ryder, B., 25

S

SAVE statement (FORTRAN 77), 6, 147
Schneiderman, B., 16
Schryer, N., 25, 27
Shampine, L., 16
sign-change, 109
Simpson, B., 149
Simpson's rule, 139–40
specification, 2
  for linear equation solvers, 94, 98
  for nonlinear equation solvers, 111
  for quadratic equation solvers, 7
  for sine procedures, 55, 67
Stewart, G., 73, 99
survival (of a mutant), 4

T

Taylor, D., 85
test, 2
  for nonlinear equation solvers, 111
  for sine procedures, 55, 65
typographical change, 3

U

Ukkonen, E., 95
ulps (units in the last place), 32–35
underflow, 19, 87
unimodal, 129
UNIX operating system, 24, 88

W

Waite, W., 21, 46, 49, 62, 67
Wang, Y., 83
Watts, H., 16
Wilkinson, J., 42–43, 73, 78, 97–98, 101–2
Wrathall, C., 83, 100, 103–4

Y

Yelowitz, L., 46

DATE DUE			

PRINTED IN U.S.A.